Strengthening
the Partnership

Improving
Military
Coordination
with Relief
Agencies and
Allies in
Humanitarian
Operations

Daniel Byman, Ian Lesser, Bruce Pirnie, Cheryl Benard, Matthew Waxman

Prepared for the United States Air Force

RAND Project AIR FORCE

This study examines how the U.S. military, particularly the U.S. Air Force (USAF), might improve cooperation with relief agencies and European allies in humanitarian relief operations. Improved coordination would help the USAF support U.S. government efforts and increase the success of relief efforts.

This study notes potential reasons for humanitarian intervention, the types of missions typically carried out by U.S. forces, and common political limits placed on military forces. With this context in mind, it describes a wide range of relief organizations, identifies problems in coordination, and offers recommendations to the USAF and to the wider military community that would improve coordination. It also examines the role of allies in relief operations and allied perspectives on key issues confronting USAF planners. The study is primarily written for military planners, but it will also interest civilians, both within and outside government, who are concerned with humanitarian assistance.

This study was sponsored by General John Jumper (USAFE/CC) and was conducted as part of the Strategy and Doctrine program of RAND's Project AIR FORCE. Comments are welcomed and may be addressed to the authors or to the Program Director, Dr. Zalmay Khalilzad.

PROJECT AIR FORCE

Project AIR FORCE, a division of RAND, is the Air Force federally funded research and development center (FFRDC) for studies and analysis. It provides the Air Force with independent analysis of policy alternatives affecting the development, employment, combat readiness, and support of current and future aerospace forces. Research is performed in four programs: Aerospace Force Development; Manpower, Personnel, and Training; Resource Management; and Strategy and Doctrine.

CONTENTS

FIGURES

TABLES

SUMMARY

Many humanitarian interventions led and supported by the United States go beyond simple disaster relief and include such difficult tasks as protecting refugees, securing humanitarian aid, and restoring civil order. Such ambitious operations—commonly referred to as "complex contingency operations"—include attempting to return a viable government to Somalia, alleviating suffering after the Rwandan genocide, and trying to create a multiethnic government in Bosnia-Herzegovina, among others. The U.S. Air Force (USAF) plays an important role in complex contingency operations as well as in smaller relief efforts. Because military support for humanitarian assistance during conflict will probably continue at a high tempo in the coming years, the USAF, and the military in general, must know and work with a wide range of actors, including U.S. government agencies, allied governments and their militaries, host nations, international organizations, and nongovernment organizations (NGOs).

Relief agencies and U.S. allies in Europe are important partners. Relief agencies usually react quickly and flexibly during crises. As a result, they are often the first on the scene and thus able to assist in assessments, relief distribution, and other vital tasks with speed and efficiency. Relief workers, many locally hired, usually know the local culture well. Some also have expertise in sanitation, disease control, nutrition, and other basic relief missions. In addition, European allies are active in complex contingency operations, and a solid partnership can yield political as well as humanitarian benefits.

This study examines how to improve coordination between the U.S. military and relief agencies during humanitarian relief operations. It also explores how the military might improve coordination with European allies in complex contingency operations. Its recommendations will help the military respond more effectively when supporting and conducting relief operations.

THE DYNAMICS OF COMPLEX CONTINGENCY OPERATIONS

Despite the frequent occurrence of crises that might justify intervention, it is difficult to predict when, where, and to what degree the United States and other major powers will conduct relief operations. Most of the civil wars, refugee flows, and other problems that in theory might lead to interventions are in countries that are remote from the United States and are not vital to U.S. security or economic interests.

In deciding to conduct operations, the United States is driven as much by domestic opinion and allied concerns as by humanitarian motives. Media coverage and grassroots efforts by NGOs can put considerable pressure on the U.S. government to act. Moved by reports of suffering abroad, local civic and church groups often collect goods—some of little immediate value in the crisis—and press local politicians to ensure they are shipped promptly. World media, however, do not report humanitarian crises consistently: Some tragedies, such as the suffering in Sudan, receive little attention, while others, such as Kosovo, received full coverage. Refugees may provide another motivation for intervention. Particularly in Europe, U.S. allies push for action when they fear massive flows of refugees into their own countries.

Military missions in complex contingency operations typically fall into five general categories: providing humanitarian assistance, protecting humanitarian assistance, assisting refugees and displaced persons, enforcing a peace agreement, and restoring order. The specific tasks necessary to carry out the missions vary widely, going far beyond standard warfighting duties. It is also common for the mission to expand or shrink in its scope or focus with little warning.

Complex contingency operations typically pose a variety of constraints on and problems for military operations that can decrease overall effectiveness. If the United States and its allies have few interests in the crisis region, they are reluctant to devote the time, resources, and attention needed to resolve the underlying political and economic problems that caused the crisis. Thus, humanitarian aid can become a substitute for political action rather than a complement to it. Intervening states seek to focus on relief, refraining from more difficult steps to stop a war or change an ineffective government. Limited U.S. and allied commitment may foster a high sensitivity to casualties, leading to restrictions on where troops can go, the types of activities conducted, and the overall rules of engagement.

The military may also have to balance political and humanitarian objectives. Both the host and donor governments often want immediate results and a visible role in providing the relief. At times, however, these motives may conflict with less-glamorous priorities, such as supplying forklifts to increase base unloading capacity. In addition, the military may be called on to collect and distribute unneeded or low-priority items to please local politicians in donor governments and their constituents. Military officials may have to work with corrupt or incompetent governments to preserve the image of partnership even when such cooperation hinders overall operations.

Complex contingency operations not only pose complex operational challenges but are also complex organizationally, involving a wide range of different and often competing or diverging actors including major powers, donor countries, host countries, international organizations, regional organizations, and NGOs. At times, everyone and no one may be in charge. Military control arrangements can be highly complex and home governments may micromanage their deployed forces. As a result, the military's mission may not be entirely clear, it may be compelled to improvise, or it may see its mission change in disconcerting ways.

AN OVERVIEW OF THE RELIEF COMMUNITY

Relief agencies differ considerably from one another. United Nations (UN) agencies play important roles in humanitarian crises, but they are highly bureaucratic and do not always coordinate well. The United Nations High Commissioner for Refugees (UNHCR) and the World Food Programme (WFP) deliver most of the assistance through UN channels. UNHCR takes the lead in caring for refugees and may act as lead agency for the entire UN relief effort in a major crisis. WFP is the world's largest multilateral provider of food aid and often runs large-scale logistics operations to deliver assistance.

NGOs form an extremely disparate group. NGOs such as Cooperative for Assistance and Relief Everywhere (CARE) and World Vision are highly professional and ready to participate in a wide range of relief activities on a global scale. At the opposite extreme, some NGOs have limited reach and consist of just a few inexperienced individuals. Some NGOs perform just one function; some represent only their country or city; and some focus on one particular area, such as the needs of children. In general, the larger and better funded organizations are also more professional and capable. But at times a small or specialized NGO may play a vital role in a particular crisis.

NGOs can be categorized into five groups for purposes of considering military-NGO coordination:

- Core-Team: organizations that are highly competent, broadly capable, and inclined to cooperate with the military. Most of them receive substantial funds through the U.S. government and are accustomed to working with government officials.

- Core-Individual: organizations that are highly competent and broadly capable, but less disposed to cooperate. In contrast to core-team NGOs, these organizations seek to avoid close ties to the U.S. government and the U.S. military, often for idealistic reasons, even though they may receive U.S. government funding.

- Specialized: organizations that are highly competent and capable in certain functional areas. In contrast to "core" NGOs, they do not provide across-the-board assistance.

- Advocacy: organizations that promote human rights or other causes but do not normally provide material assistance.

- Minor: organizations that may or may not be competent and are less capable than the core organizations.

This typology helps clarify the most important actors and allows the military to apportion its resources effectively.

BARRIERS TO BETTER COOPERATION

Coordination between the military and the relief community is uneven and constrained for a variety of reasons. Agencies in the UN family in particular suffer from several limitations. By comparison with NGOs, most UN agencies move slowly. Because UN agencies are highly autonomous, interagency coordination is often faulty. Finally, they are compelled by the UN Charter to work closely with host governments (which are normally member states in the organization), rather than directly with local populations. This close relationship may corrupt or divert the flow of assistance.

UN and NGO time horizons may vary from those in the military. Personnel may be in the area before the military arrives and remain after it departs. They often see the military as an expensive, flashy, and sometimes disruptive interloper that will accomplish a few well-publicized tasks and depart suddenly. They have little sympathy with the military's concern for exit strategy because they believe that real improvement requires a long-term commitment.

NGOs and the military have different organizational cultures. Most NGOs are accustomed to decentralized decisionmaking and tend to be scornful of military hierarchies. Most NGOs plan poorly or not at all, and they may doubt whether the exertion of force can effect any lasting improvement. Many NGOs are reluctant to accept protection until it becomes a necessity and even then tend to ignore rules. They are critical of the military's penchant for classifying information, even information that is openly available. In recent years, however, mutual understanding and appreciation have grown, leading both the military and the relief agencies to seek closer cooperation.

One difference that cannot be overcome is the desire to maintain neutrality and impartiality. Many NGOs are committed to providing assistance on the basis of need without regard to politics. Moreover, their reputation for neutrality and impartiality is their best protection. To move freely in an area of conflict and provide assistance to all victims, NGOs must convince combatants that they will not assist any side preferentially. NGOs may eschew close or open ties to the military, if such ties would compromise their reputation and open them to reprisals.

Military personnel and NGO workers often have little understanding of each other's organizations and procedures. Most military officers have only a limited knowledge of NGOs and cannot distinguish major organizations from minor ones. NGO workers may be ignorant of the military and have unrealistic expectations of what the military can provide.

NGOs may also doubt the U.S. government's willingness to commit its military to humanitarian missions. Many NGOs, believing that the United States approaches humanitarian relief in an ad hoc manner, hesitate to devote resources to improving ties to the military because the military may withdraw abruptly during a crisis or not help at all.

THE EUROPEAN CONTRIBUTION

European allies have a long tradition of humanitarian intervention, especially on the periphery of Europe. Humanitarian missions are becoming a central focus of European defense planning on a national and European Union (EU) level. European allies possess substantial tactical lift assets and are relatively active in using them to support relief operations in Africa and the Balkans. They are also more inclined than is the United States to donate military airlift for relief operations. France (and to a lesser extent, Belgium) is in the first rank of such operations in Africa. A second-tier group of moderately active states includes Italy and the United Kingdom. A third, but still active, grouping includes Portugal, Germany, the Netherlands, Canada, and Spain.

Linguistic, cultural, and political-military ties to former colonial areas lend some advantages (and some disadvantages) to European allies operating in Africa and elsewhere. The style of European humanitarian intervention, both military and NGO, differs from that of the United States. The scale is smaller, the engagement less distant, and the emphasis on force protection is noticeably less evident. Participants are often more familiar with local actors and conditions.

In large-scale humanitarian relief operations, as in Central Africa, Somalia, and the Balkans, there is a potential synergy between European and U.S. capabilities. European allies have long-standing local ties, useful tactical lift, and extensive training for humanitarian missions. The United States can offer heavier lift to the theater; big-picture and near–real time intelligence; and command, control, and communications (C3) assets, without which large-scale operations may be delayed or ineffective. Not least, European allies, through the EU, are in a position to provide substantial funding for humanitarian relief.

Post-Kosovo, NATO has emerged as a key actor and stakeholder in humanitarian crises. NATO structures and capabilities in civil-emergency planning, logistics, and the coordination of operations give the Alliance a special role in complex humanitarian missions. The new Strategic Concept, and Balkan precedents, suggest that NATO will be an increasingly important actor in this sphere.

RECOMMENDATIONS FOR THE U.S. MILITARY

To improve coordination in future crises, the military—together with other interested agencies in the U.S. government—should consider several steps. At a minimum, the military should ensure that key personnel are broadly familiar with those organizations that are most relevant to humanitarian relief operations. Officers in unified commands should have responsibility for identifying major organizations, especially NGOs, in the command's area of responsibility and maintaining regular contact with them. At the same time, the unified commands and the services should help familiarize these organizations with the military's organization and capabilities. Relevant organizations would include agencies in the United Nations family, the International Committee of the Red Cross (ICRC), and the core-team NGOs. Mutual familiarization would promote mutual understanding and better cooperation. To help improve familiarity, the unified commands should consider appointing a humanitarian advisor (comparable to a political advisor) and expanding their relationship with the Center of Excellence that is currently working with U.S. Pacific Command (USPACOM).

In addition, the military should try to improve long-term planning and coordination by engaging the most important relief agencies and bringing them into the planning process. Key organizations would include several agencies in the United Nations family (e.g., WFP and UNHCR), the ICRC, and a selection of NGOs—especially the core-team NGOs. The unified commands should establish regular contact with officials at these key agencies, inviting them to play roles in the planning process and, in cooperation with other U.S. government elements, encouraging them to develop relief packages that could be quickly deployed during crises. The services and the unified commands should regularly consult officials from these organizations before and during a crisis and transport their personnel into the area when necessary. Closer ties to the key agencies would speed response and increase efficiency during all phases of a humanitarian crisis, but they would pay particular dividends during the initial phase of operations, when delay can cost lives. Most NGOs would respond favorably to this selective approach, although some might allege favoritism if they were not consulted and others might avoid contacts with the military to preserve a reputation for impartiality. To minimize this problem, the military should work through umbrella organizations and civilian government agencies.

Exercises should also involve the core-team and other major NGOs more extensively. NGO personnel should participate in planning appropriate portions of exercise scenarios and be allowed free play to the extent possible. Exercises should include realistic play of vital aspects of relief operations such as managing airflow, establishing a Civil-Military Operations Center (CMOC), and developing procedures to protect relief agencies from banditry and looting.

The military should encourage efforts to improve information sharing. It should identify NGOs with on-the-ground networks and promote information exchanges with them. The military should minimize classification of data that should be shared among military and civilian actors. It should also share after-action reports with relief agencies and solicit their responses.

Building on both these efforts, the military should initiate actions to improve coordination of the relief flow during humanitarian crises. All the services and unified commands could offer their logistics expertise, which is often lacking among NGOs and UN agencies, in the early days of the crisis. The services could manage the overall relief coordination effort until NGOs and UN agencies had the personnel and expertise in place to take over this function.

Airlift is particularly important. Only the USAF has the capacity to quickly conduct a massive airlift, which is often necessary early in a crisis. To be fully effective, the USAF and the unified commands should address both the narrower problem of air traffic control and the more fundamental problem of establishing priorities. The USAF can provide the capacity by itself, but it must work with other U.S. government elements to convince NGOs to employ the capacity and to establish relief priorities.

Given the growing role of European allies and changes in the involvement of NATO and other key organizations, there are now worthwhile opportunities to improve cooperation with allies in humanitarian contingencies. Where possible, the United States should strengthen NATO's capacity for civil-emergency planning and humanitarian relief and consider creating a NATO Assistant Secretary General for Civil Emergency Planning. Washington should also place transatlantic cooperation in planning for humanitarian response high on the prospective NATO-EU agenda.

Operational steps would also help improve coordination with allies. The United States should explore arrangements to take advantage of French facilities and European relationships in and around Africa to support relief operations. Washington should take advantage of allied interest to promote inter-operability in humanitarian airlift and airdrop. Finally, the United States should provide places for European NGOs at relevant U.S. and NATO courses and war colleges.

On its own, the military cannot solve the coordination problems inherent in humanitarian assistance. A more complete solution would require the efforts of many actors, including major donor countries and host countries at high political levels. But the military can advocate more comprehensive action while working on those aspects of the problem that fall within its sphere of responsibility. Even within its own sphere, the military can achieve considerable improvement and act as a catalyst for broader reform.

ACKNOWLEDGMENTS

The authors gratefully acknowledge the help of many colleagues and interlocutors inside the government and among relief agencies. Particular thanks go to Major Karen Kwiatkowski, Major Thomas Headen, and Lt. Colonel John Williams, who provided valuable input into our project and assisted in arranging interviews, acquiring research documents, and otherwise helping us gather data. Michael Hix of RAND provided a superb and detailed critique that greatly strengthened the analysis in this report.

Our research required a large number of interviews with aid organization personnel. These individuals provided insight into all aspects of our research, and we greatly appreciate their time and candor. They include Alan C. Alemian, John Ashton, Gaspar F. Colon, Donna Derr, Walter Franciscovich, Antoine Gerard, Kenneth Hackett, Amy Hilleboe, Larry Hollingsworth, Elton F. King, Terry Kirsch, Lauren Landis, Jerry Martone, Dayton Maxwell, Kim Maynard, Robert McPherson, Daniel N. Mushala, Andrew Natsios, Sister Maura O'Donohue, Daniel Philippin, Annemarie Reilly, and Jean-François Vidal.

United Nations officials, and those of affiliated agencies, also provided fresh insights and a clear understanding of the problems and limits faced by various international organizations. Thanks go to Karen Koning Abuzayd, Dawn T. Calabia, Major General (ret.) Wilfried De Brouwer, Michael Elmquist, Kevin M. Kennedy, David M. Kirkham, Guillaume de Montravel, and Jean-Daniel Tauxe.

This study benefited from the insights and expertise of many U.S. military and defense department officials. They include Colonel (S)

Hank Andrews, Lt. Colonel Paul Cariker, Lt. Colonel Alan Cox, Lt. Colonel G. Michael Dudley, Colonel James Fellows, Master Sergeant Donald W. Gripp, David W. Hamon, Colonel Michael Hess, Colonel Paul D. Hughes, Lt. Commander Mark Laxen, Albert Mitchum, Linda Mcree, Colonel Jose Negron, Suellen B. Raycraft, Deborah Rosenblum, James Schear, Captain Chris Seiple, Major Kenneth W. Shreves, Lt. Colonel Kevin B. Smith, Lt. Colonel Steven J. Thompson, Nancy Walker, and Master Sergeant David Webb.

Other U.S. government officials who contributed include Ted Constantine, Paul Frandano, Ellen Laipson, Ambassador Hugh Montgomery, Jeremy Reiskin, Richard Ross, Enid C.B. Schoettle, and John Sullivan.

This research benefited greatly from discussions with officials and unofficial observers in the United States and Europe, including individuals at the NATO International Staff, the U.S. Mission to NATO, and SHAPE. Given the importance of the French role in any discussion of allied contributions to relief operations, we are especially grateful to Natalie Fustier and Bruno Tertrais for their assistance in arranging interviews at the French Ministry of Defense, the headquarters of the Force Aerienne de Projection, and the International Operations Office of the French Red Cross.

Several experts in the academic community gave freely of their time and expertise, helping to sharpen our conclusions and provide a better framework for understanding. Thanks go to Manuel Carballo, Antonia Handler Chayes, Patrick Doherty, John Hammock, James Kunder, and Taylor Seybolt.

People in the Center of Excellence in Disaster Management & Humanitarian Assistance located at Tripler Army Medical Center, Hawaii, gave generously of their time during our visit. We thank Enzo Bollettino; Frederick ("Skip") M. Burkle, Jr.; David G. Haut; Robin Hayden; Gary J. Rhyne; James H. Rogers; and Mark S. Schmidt.

Several U.S. government, military, and aid organization officials asked not to be named. Their anonymity should in no way diminish their contributions to this project.

Several RAND colleagues provided invaluable assistance in the research and formulation of this project. Paul Killingsworth assisted in obtaining data and keeping us abreast of related developments in the Air Force. Zalmay Khalilzad and Alan Vick offered overall direction for the project with their comments and suggestions. Jennifer Ingersoll Casey compiled the material presented in Appendices A and B. Thanks also go to Donna Boykin for her administrative assistance.

ABBREVIATIONS AND ACRONYMS

AAI	African-American Institute
ACDI/VOCA	Agricultural Cooperative Development International / Voluntary Overseas Cooperative Assistance
ACF	Action Against Hunger
ADRA	Adventist Development and Relief Agency
AJWS	American Jewish World Service
AMC	Air Mobility Command
AMCC	Allied Movement Coordination Center
AME	Air Mobility Element
AOR	Area of responsibility
ARC	American Refugee Committee
ASG	Assistant Secretary General
BBF	The Brother's Brother Foundation
C3	Command, Control, Communications
CARE	Cooperative for Assistance and Relief Everywhere
CCF	Christian Children's Fund
CDC	Center for Disease Control
CENTAUR	Combined Event Notification Technology and Unified Reporting

CFSP	Common Foreign and Security Policy
CHART	Combined Humanitarian Assistance Response Training
CIMIC	Civil Military Cooperation Cell
CINC	Commander-in-Chief
CINCPAC	Commander-in-Chief, U.S. Pacific Command
CIS	Commonwealth of Independent States
CJCIMIC	Combined Joint Civil Military Cooperation
CJCMTF	Combined Joint Civil-Military Task Force
CJCS	Chairman, Joint Chiefs of Staff
CJTF	Combined Joint Task Force
CMMB	Catholic Medical Mission Board
CMOC	Civil-Military Operations Center
COE	Center of Excellence
CPP	Cambodian People's Party
CRG	Contingency Response Group
CRS	Catholic Relief Services
CWS	Church World Service
DART	Disaster Assistance Response Team
DHA	Department of Humanitarian Affairs
DPK	Democratic Party of Kurdistan
DPKO	Department of Peacekeeping Operations
EADRCC	Euro-Atlantic Disaster Response Coordination Center
EAPC	Euro-Atlantic Partnership Council
ECHO	European Community Humanitarian Office
ECOMOG	ECOWAS Cease-Fire Monitoring Group

ECOSOC	Economic and Social Council
ECOWAS	Economic Community of West African States
ERC	Emergency Relief Coordinator
EU	European Union
FAO	Food and Agricultural Organization
FAP	Force Aerienne de Projection
FAR	Forces Armees Rwandaises
FHI	Food for the Hungry International
FRETILIN	Frente Revolucionaria do Timor Leste Independente
FUNCINPEC	United Front for an Independent, Neutral, Peaceful, and Cooperative Cambodia
HELP	Health Emergencies in Large Populations
HPI	Heifer Project International
HUMAD	Humanitarian Affairs Advisor
IASC	Inter-Agency Standing Committee
ICITAP	International Criminal Investigative Training Assistance Program
ICRC	International Committee of the Red Cross
ICRW	International Center for Research on Women
ICTR	International Criminal Tribunal for Rwanda
ICVA	International Council of Voluntary Organizations
IDP	Internally displaced persons
IFOR	Implementation Force, Operation Joint Endeavor
InterAction	American Council for Voluntary International Action
IO	International organization
IQC	Indefinite Quantity Contract

IRC	International Rescue Committee
IRSA	Immigration and Refugee Services of America
JTF	Joint Task Force
KFOR	Kosovo Force
KLA	Kosovo Liberation Army
MCD	Medical Care Development
MCDU	Military and Civil Defense Unit
MINUGUA	(UN) Mission for the Verification of Human Rights in Guatemala
MINURCA	United Nations Mission in the Central African Republic
MIPONUH	United Nations Civil Police Mission in Haiti
MISAB	Mission Interafricaine de Surveillance des Accords de Bangui
MONUA	United Nations Observer Mission in Angola
MOOTW	Military operations other than war
MPLA	Popular Movement for the Liberation of Angola
MRTA	Movimiento Revolucionario Tupac Amaru
MSF	Médecins Sans Frontières (Doctors Without Borders)
NAC	North Atlantic Council
NADC	NATO Air Defence Committee
NATO	North Atlantic Treaty Organization
NCA	National command authority
NGO	Nongovernment organization
OAU	Organization of African Unity
OCHA	Office for the Coordination of Humanitarian Affairs

OFDA	Office of Foreign Disaster Assistance
OHR	Office of the High Represensative
OPM	Organisasi Papua Merdeka (Free Papua Movement)
OSCE	Organization for Security and Cooperation in Europe
Oxfam	Oxford Committee for Famine Relief
PAM	Programme Alimentaire Mondial
PDD	Presidential Decision Directive
PDK	Party of Democratic Kampuchea
PDMIN	Pacific Disaster Management Network
PFP	Partnership for Peace
PHR	Physicians for Human Rights
PIC	Peace Implementation Council
PKK	Kurdish Workers Party
PUK	Patriotic Union of Kurdistan
PVO	Private voluntary organization
RI	Refugees International
ROE	Rules of engagement
RPF	Rwandese Patriotic Front
RSCC	Refugee Support Coordination Center
RUF	Revolutionary United Front
SACEUR	Supreme Allied Commander, Europe
SACLANT	Supreme Allied Commander, Atlantic
SAM	Surface-to-air missile
SAT	Southern Air Transport
SHAPE	Supreme Headquarters Allied Powers Europe

SLA	South Lebanon Army
SFOR	Stabilization Force
TALCE	Tanker Airlift Control Element
UMCOR	United Methodist Committee on Relief
UN	United Nations
UNAMIR	United Nations Assistance Mission for Rwanda
UNAMSIL	United Nations Mission in Sierra Leone
UNAVEM	United Nations Angola Verification Mission
UNDP	United Nations Development Programme
UNFICYP	United Nations Peacekeeping Force in Cyprus
UNHCR	United Nations High Commissioner for Refugees
UNICEF	United Nations Children's Fund
UNIFIL	United Nations Interim Force in Lebanon
UNITA	União Nacional para a Independência Total de Angola (National Union for the Complete Independence of Angola)
UNITAF	Unified Task Force
UNMIBH	United Nations Mission in Bosnia and Herzegovina
UNMIK	United Nations Interim Administration in Kosovo
UNMOGIP	United Nations Military Observer Group in India and Pakistan
UNMOT	United Nations Mission of Observers in Tajikistan
UNOCHA	United Nations Office for the Coordination of Humanitarian Assistance to Afghanistan
UNOMIG	United Nations Observer Mission in Georgia
UNOMIL	United Nations Observer Mission in Liberia

UNOMSIL	United Nations Observer Mission in Sierra Leone
UNOSOM I	First United Nations Operation in Somalia
UNOSOM II	Second United Nations Operation in Somalia
UNPREDEP	United Nations Preventive Deployment Force
UNPROFOR	United Nations Protection Force
UNTAC	United Nations Transitional Authority in Cambodia
USACOM	U. S. Atlantic Command
USAF	U.S. Air Force
USAFE	U.S. Air Forces in Europe
USAID	U.S. Agency for International Development
USCC	United States Catholic Conference
USCENTCOM	U.S. Central Command
USCR	United States Committee for Refugees
USEUCOM	United States European Command
USG	United States Government
USPACOM	U.S. Pacific Command
USSOUTHCOM	U.S. Southern Command
USTRANSCOM	U.S. Transportation Command
VOICE	Voluntary Organizations in Cooperation in Emergency
WEU	Western European Union
WFP	World Food Programme
WHO	World Health Organization
WVRD	World Vision Relief and Development
YMCA	Young Men's Christian Association

Chapter One

INTRODUCTION

The U.S. military has always supported relief operations following natural disasters, but since the end of the Cold War it has increasingly assisted the victims of conflict. In recent years, the military has conducted major operations in northern Iraq, Turkey, Somalia, Haiti, Rwanda, Zaire, Bosnia, and Kosovo. In four of these cases (Somalia, Haiti, Bosnia, and Kosovo), the U.S. military was also heavily involved in peace operations. The U.S. Air Force (USAF) supported all of these operations and many smaller ones as well.

The USAF plays a vital role in conducting these operations.[1] USAF lift assets are regularly employed in relief operations, ranging from a few flights of food to a massive airlift to sustain refugees. Beyond the provision of lift, the USAF also plays an essential role in supporting other services and as part of the unified command's overall relief effort. The USAF assists in providing intelligence, deterring potential combatants, and managing the overall relief effort.

Relief operations to aid victims of man-made disasters are both organizationally and politically complex. If not conducted with great care, they may even increase human suffering by provisioning combatants and thus fueling a conflict. For example, the primary recipients of aid during Support Hope, the U.S. operation in Zaire (now Congo) following the Rwandan genocide, were Hutu refugees, many still organized and controlled by a genocidal leadership. This leadership intended to return to Rwanda by force and conducted

---

[1]For an overview of USAF participation in military operations other than war (MOOTW), see Vick et al. (1997).

bloody raids from refugee camps in Zaire but was crushed in a Tutsi-led invasion of eastern Zaire.

Effective operations demand coordination among a wide range of actors that include civilian departments of the U.S. government, especially the State Department and U.S. Agency for International Development (USAID); international organizations, especially agencies of the United Nations; and nongovernment organizations (NGOs).[2] As almost all operations are multilateral, working with allies is also essential, particularly in Europe.   Coordination is essential to deliver and sustain relief operations in the most effective manner.   Coordinating these entities, however, is a daunting task. The interagency process within the U.S. government often functions inadequately during humanitarian crises, leading to incoherent or unclear formulations of strategy.[3] The United Nations is composed of numerous, often-competing agencies whose lines of authority and areas of responsibility overlap. NGOs are extremely numerous— several thousand NGOs have consultative status with the Economic and Social Council of the United Nations[4]—and are not subject to central direction.

This study examines ways to improve coordination between relief agencies and the military during relief operations. A caveat is in order.   The term *military* is imprecise at best. We have used it to include the individual services in their Title X capacity, the unified commands, and the military departments. When appropriate, we identify the specific entity within the military; when the issue encompasses all these entities, we use only the term *military*.

Although the report describes problems shared by the military as a whole, the focus of the recommendations is on the unified commands and individual services, both in their Title X capacity and

---

[2]According to joint doctrine, a nongovernment organization is "a transnational, non-profit organization of private citizens that maintains a consultative status with the Economic and Social Council of the United Nations. Nongovernment organizations may be professional associations, foundations, multinational businesses, or simply groups with a common interest in humanitarian assistance (development and relief)." Joint Chief of Staff (1966), Joint Pub 3-08, p. I-15, footnote 3. A "private voluntary organization" (PVO) is an NGO that is properly registered with the USAID.

[3]Pirnie (1998).

[4]United Nations (1998), p. 13.

in their support for the unified commands. Because of the sponsorship of this report, we focus more on the USAF than on other services, but almost all the problems we identify cannot be fixed by single-service responses: They must be addressed in a joint context or by the broader military community as a whole. Thus, at times we also identify policies that the joint staff and civilian agencies should consider in order to implement the changes we recommend for the services and the unified commands. We also recommend actions at higher political levels within the U.S. government when we suggest ways to work better with European allies.

Successfully providing relief, of course, requires far more than better coordination, but this is an essential step. Although the implications of this study are relevant to a range of military operations other than war (MOOTW) concerns, this study emphasizes coordination during relief operations to aid victims of man-made disasters, usually conducted in risky environments. Examples include Provide Comfort I (Turkey and Northern Iraq, 1991), Restore Hope (Somalia, 1992–1993), and Support Hope (Rwanda, Zaire 1994).[5]

This study of how to improve cooperation is directed at four audiences: (1) military planners and operators, particularly in the USAF and unified commands, who are tasked to conduct these operations; (2) U.S. government policymakers trying to improve the overall response to crises; (3) UN officials and NGO personnel who work with the military; and (4) members of the general public concerned with humanitarian assistance.

RESEARCH APPROACH AND STRUCTURE

This report drew on three basic sources for data: (1) current literature on relief operations, (2) NGO reporting, and (3) interviews. The authors used after-action reports, case studies, journalistic accounts, and academic studies of particular interventions and on the general

---

[5]This study does not focus on operations conducted to aid victims of natural disasters, usually conducted in a benign security environment, such as Sea Angel (Bangladesh, 1991) and Hurricane Mitch (Central America, 1998–1999). It does not directly address peace operations, such as Joint Endeavor (former Yugoslavia, 1995–1996), intended to enforce the implementation of agreements. Its conclusions, however, apply more generally to all relief operations.

subject of humanitarian relief. To learn about NGO capabilities and funding, the authors relied on materials presented by NGOs to the U.S. government, their membership, and the general public. Finally and most importantly, the authors conducted interviews, usually in person, with U.S. government officials, allied government officials, U.S. and allied military officers, UN officials, NGO personnel, and academic experts. These interviews were especially helpful in understanding current difficulties, assessing the usefulness of improved cooperation, and framing recommendations.

This report has four parts. Part One examines a particular type of intervention, commonly called complex contingency operations, that the U.S. military is often called on to perform today. It describes the frequency of conflict, military tasks, and the implications for coordination with relief partners. Part Two examines in more detail the question of military–relief agency cooperation in complex emergencies. It provides an overview of the relief community and notes obstacles to cooperation. Part Three assesses the role of European allies, noting their perspective on complex contingencies and what they can bring to the table. The report concludes in Part Four by suggesting ways the military can improve its coordination with relief agencies and with European allies. The focus of these recommendations is not only on the USAF, but also on what the other services, unified commands, Defense Department, and the U.S. government can do to better improve coordination.

PART ONE. COMPLEX CONTINGENCY OPERATIONS AND THE ROLE OF THE MILITARY

Since the end of the Cold War, the United States has led several major relief operations that often drew together its own efforts and those of host countries, neighboring states, European allies, international organizations, regional organizations, and NGOs to achieve ambitious goals. In many of these operations, U.S. activities extended beyond disaster relief and included such difficult tasks as reconstituting a government, caring for refugees, or stopping ethnic terror—operations generally referred to as *complex contingency operations.*

This section first describes complex contingency operations, discusses current conflicts, and then notes common tasks assigned to the military in these operations. It concludes by assessing how complex contingency operations pose challenges for military operations. Together, these four chapters set the stage for the subsequent discussion of how the military can improve its performance in these operations through better cooperation with relief agencies and with allies.

Chapter Two

CHARACTERISTICS OF COMPLEX CONTINGENCY OPERATIONS

Complex contingency operations differ from such typical relief operations as caring for victims of an earthquake or providing support to an allied government suffering a crop failure. Complex contingency operations are both more difficult and more demanding. They often require balancing uncertain domestic support, differing allied goals, varying bureaucratic interests, and other political factors, often for a considerable length of time. Moreover, the military is regularly called upon to perform difficult and unusual tasks, such as separating combatants and providing for refugees, that are not necessary in more typical disaster environments. Indeed, unlike relief after a natural disaster, the provision of relief in response to a civil war or the depredations of a brutal government can strengthen combatants and actually worsen a conflict.[1]

The United States and its allies have recently conducted a wide range of complex contingency operations. Examples of recent operations include a failed attempt to reconstitute viable central government in Somalia (Operation Restore Hope, Operation Continue Hope), return of democratically elected government to Haiti (Operation Uphold Democracy), alleviation of human suffering in Rwanda and Zaire (Operation Support Hope), operations to end conflict and recreate multiethnic government in Bosnia-Herzegovina (Operation Joint

---

[1]For this report, we use the term *relief operations* when referring to a broad category that encompasses complex contingency operations, which are the most difficult of relief operations. Although the focus of our work is on complex contingencies, most of our arguments apply to relief operations in general.

Endeavor, Operation Joint Guard), and an effort to stop ethnic terror in Kosovo (Operation Allied Force, Operation Shining Hope, and Operation Joint Guardian).

These types of operations go beyond simple disaster relief, demanding the coordination of multiple actors. Moreover, they often require a response to man-made crises such as civil war or poor governance in addition to alleviating humanitarian disasters. Officially, they are called complex contingency operations. Presidential Decision Directive-56 (PDD-56) defines the term complex contingency operations by examples:

> . . . peace operations such as the peace accord implementation operation conducted by NATO in Bosnia (1995–present) and the humanitarian intervention in North Iraq called Operation Provide Comfort (1991); and foreign humanitarian assistance operations, such as Operation Support Hope in Central Africa (1994) and Operation Sea Angel in Bangladesh (1991).[2]

As the above examples suggest, complex contingencies can be understood as much by what they are not as by what they are. The term does not include smaller operations such as domestic disaster relief, counterterrorism, hostage rescue, and noncombatant evacuation. Nor does it include international armed conflict at the other extreme.

Complex contingency operations may be categorized by certain fundamental decision points and the implied branches as depicted in Figure 2.1. The United States must first decide whether the operation is simple, implying that PDD-56 does not apply, or complex, implying that PDD-56 does apply and a political-military plan should be prepared. The United States must next decide whether to use military force coercively or not to do so. If the United Nations is involved, as it usually is, noncoercive operations imply that Chapter VI of its charter will be invoked, whereas coercive operations fall under Chapter VII. Finally, U.S. decisionmakers must decide whether to aid victims of conflict without trying to resolve the conflict that caused their suffering or to seek resolution, usually

---

[2]U.S. Government, *The Clinton Administration's Policy on Managing Complex Contingency Operations*, PDD-56 (May 1997).

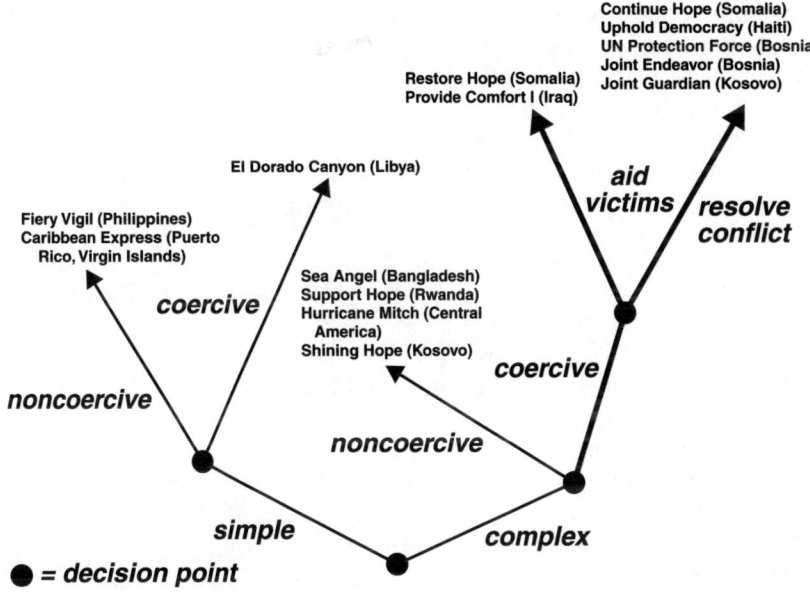

Figure 2.1—Contingency Operations

implying implementation of a peace plan. Each category has its own risks and requirements.

Simple noncoercive contingencies are usually relief operations following a natural disaster, such as the eruption of Mount Pinatubo in the Philippines (Operation Fiery Vigil) or Hurricane Marilyn in Puerto Rico and the Virgin Islands (Operation Caribbean Express). Simple coercive operations include shows of force, air denial operations, and air strikes, such as the strike against Libya during the Reagan administration (Operation El Dorado Canyon). As Operation El Dorado Canyon illustrates, these operations may entail considerable operational and tactical complexity, but they are simple in the sense that one actor leads and performs most of the effort.

Complex noncoercive operations may respond to natural disasters, such as Typhoon Marian in Bangladesh (Operation Sea Angel) or Hurricane Mitch in Central America. PDD-56 explicitly identifies

Operation Sea Angel as a complex contingency operation.[3]   Such operations may also respond to situations in which conflict is involved, but if conflict is the underlying problem, a decision not to coerce may have undesirable consequences. For example, Operation Support Hope saved lives and did not involve coercion on the part of U.S. troops, but it also inadvertently nourished a genocidal Hutu regime that dominated the refugee camps in Zaire.

Complex coercive operations are at the high end of the scale of contingencies. They generally have two main goals: (1) provide humanitarian assistance without trying to resolve conflict and (2) try to resolve conflict, typically through enforcement of a peace plan. (The second goal could, of course, include humanitarian assistance as well.) The first alternative has limitations that have been dramatically apparent in practice. For example, Operation Restore Hope successfully secured aid in a Somalia torn by factional warfare, but it led directly to the disastrous failure of Operation Continue Hope, an operation launched with more ambitious goals but less effective forces. Likewise, Operation Provide Comfort helped Iraqi Kurds return home, but it left the underlying problems of Iraqi oppression and Kurdish factionalism unresolved. As a result, pro-U.S. Kurds were later compelled to flee and the United States still mounts a seemingly interminable air denial operation (Operation Northern Watch).[4]

---

[3]U.S. Southern Command (USSOUTHCOM) considered Hurricane Mitch to be a complex contingency that should have triggered PDD-56 procedures but did not.

[4]Coercive operations to resolve conflict are usually well publicized and highly controversial, posing difficult decisions for U.S. and allied leaders. Operation Continue Hope was intended to enforce implementation of the Addis Ababa Agreements among the Somali factions, but it depended on U.S. leadership and failed when the United States withdrew. Operation Uphold Democracy restored democratically elected government to Haiti, but it remains to be seen whether Haitians can perpetuate democratic practice despite a lack of a democratic tradition and grinding poverty. The United Nations Protection Force (UNPROFOR) had Chapter VII authority, but failed to accomplish its tasks, especially protection of "safe areas" such as Bihac, Srebrenica, and Zepa. Indeed, UNPROFOR could not even protect its own members, whom the Bosnian Serbs seized as hostages to avert NATO air strikes. NATO operations (Operation Joint Endeavor, Operation Joint Guard) have successfully enforced the military provisions of the Dayton Agreements, but the parties continue to resist implementation of the political provisions, despite pressure from the High Representative. During Operation Allied Force, NATO also carried out Operation Shining Hope to assist Kosovar refugees.

These categories blur in practice. Shining Hope was strictly a relief operation, but it was conducted in conjunction with the coercive Allied Force. Support Hope relied on military force for protection even though the main emphasis was on assisting the provision of relief. Nevertheless, this typology illustrates key decision points and clarifies the subsequent discussion of the types of missions and corresponding difficulties.

Chapter Three

FREQUENCY OF CONFLICT AND RESPONSE

There is no dearth of conflicts that might trigger complex contingency operations today. Refugee flows, communal strife, state-initiated famines, and other causes of suffering will probably be all too common in the coming years. Certainly in the Balkans and sub-Saharan Africa, but probably also in central and southeastern Asia, bellicose national and ethnic leaders may cause human suffering on a scale that might prompt and justify future U.S. intervention.

Most of these conflicts either occur within an existing state or have a strong intrastate component. The fundamental causes of conflict are many and varied, but national and ethnic antagonisms predominate. Conflict will remain prevalent in the area of responsibility of the United States European Command (USEUCOM), especially in sub-Saharan Africa but also to a lesser extent in the former Yugoslavia.

This chapter describes the current state of conflict in the world today. It lists overall conflicts worldwide and focuses on the former Yugoslavia and sub-Saharan Africa, which appear particularly prone to unrest. It then notes why it is difficult to predict U.S. and allied intervention despite the regularity of conflict.

INCIDENCE OF CONFLICT WORLDWIDE

Both interstate and intrastate conflict can prompt a humanitarian intervention that may be characterized as a complex contingency. Interstate conflict occurs infrequently, but there are several potential flash points. These include war between the Republic of Korea and a decaying Communist regime in the north, war between India and

Pakistan (possibly over Kashmir), and continuing risk of aggression from a totalitarian Iraq. Africa is home to several deadly wars. Eritrea and Ethiopia recently terminated a border conflict that cost some tens of thousands of lives. Rwanda, Uganda, Zimbabwe, and Angola are involved in a proxy war in Congo, which shows little sign of abating.

In contrast, intrastate conflict occurs often, although it has become slightly less prevalent in recent years.[1] Intrastate conflict usually has deep underlying causes. These include oppressive, illegitimate, or incompetent governments; dramatic class differences and economic grievances; ideological and religious antagonisms; national, racial, and ethnic differences; and gang-style rivalries to exploit sources of wealth.[2]

The most common underlying cause of intrastate conflict is national, racial, and ethnic differences, often sharpened or expressed by religious divides, as displayed in Table 3.1. Many people, especially in the Balkans and sub-Saharan Africa, define themselves primarily as belonging to some group with a shared past and common destiny. They intend government to advance their group's well-being, rather than to safeguard the rights of all citizens. Where one group predominates, it may abuse state power to dispossess and terrorize minorities, as Serbs did recently to Kosovar Albanians. It may even commit genocide, as Hutus did to Tutsis in Rwanda.

One symptom of conflict is people driven from their homes, often under conditions of great misery. The United Nations High Commissioner for Refugees (UNHCR) originally had a mandate to protect and assist refugees, i.e., people who fled war or persecution across international borders. But in response to intrastate conflicts, UNHCR has widened its mandate to include "internally displaced persons," i.e., those forced from their homes within the borders of recognized states. The number of persons of concern to UNHCR reached a record high of 27 million in 1995 and has since dropped to 22.3 million. Of these persons, 7.3 million are in Africa, 7.4 million in Asia, and 6 million in Europe. Of the ten largest refugee populations,

---

[1] Khalilzad and Lesser (1998); Wallensteen (1998).

[2] Byman and Van Evera (1998).

Table 3.1

Recent and Current Intrastate Conflicts

Country	Character of Conflict	International Response
Afghanistan[a]	Uzbek- and Tajik-based opposition to the Taliban regime in Kabul.	
Algeria	Violent Islamist insurgency led by Islamic Salvation Front and Armed Islamic Group.	
Angola[a]	Protracted civil war between an MPLA-dominated government and UNITA.	(Previously: UNAVEM I, II, III, and MONUA)
Azerbaijan[a]	Armenian revolt in Nagornyi Karabakh against Azeri-dominated government.	Mediation by OSCE.
Bosnia-Herzegovina[a]	Postwar tension among Bosnians, Croats, and Serbs; recalcitrance in implementing Dayton Agreement; sporadic violence provoked by minority returns.	Dayton Agreements implemented and enforced by UNMIBH, OSCE, PIC, NATO-led SFOR.
Burundi[a]	Tutsi-dominated government confronts restive Hutu majority.	OAU mediation and pressure.
Cambodia	Tension among CPP, FUNCINPEC, and PDK.	(Previously: UNTAC)
Central African Republic	Repeated Army rebellions against civilian government.	MINURCA; French-supported MISAB.
Chad[a]	Insurgency following a civil war between Muslim and Christian factions together with a Libyan invasion opposed by France.	
Colombia	Government fighting insurgent groups and drug cartels.	
Congo-Brazzaville	Tension following a destructive civil war among party militias and factions within the armed forces.	
Congo-Kinshasa[a]	Tension following civil war; weak central government; ethnic rivalries.	
Cyprus[a]	Tension between Greek Cypriots and Turkish Cypriots divided by a buffer zone; corresponding Greek-Turkish tension.	UNFICYP; UN-sponsored negotiations involving U.S. Special Envoy.

Table 3.1—continued

Country	Character of Conflict	International Response
Ethiopia[a]	Tensions among ethnic groups; drought; conflict with Eritrea.	
Georgia[a]	Separatist movement in Abkhazia.	UNOMIG; CIS peacekeeping force.
Guatemala	Tension following protracted civil war.	MINUGUA.
Guinea-Bissau	Conflict between government and military junta.	ECOMOG.
Haiti	Inadequate governance; political violence.	MIPONUH.
India[a]	Muslim opposition in Jammu and Kashmir; recurrent Hindu-Muslim violence.	UNMOGIP.
Indonesia[a]	Islamist opposition in Java; FRETILIN opposition in East Timor; OPM opposition in Irian Jaya; class and racial antagonisms.	
Iraq[a]	Kurdish opposition led by DPK and PUK (DPK currently aligned with government).	Northern Watch.
Lebanon[a]	Sporadic combat and terrorism involving Israeli forces, SLA, Hezbollah, and Syrian forces.	UNIFIL.
Liberia[a]	Tension following a protracted civil war among ethnically based factions.	(Previously: UNOMIL; ECOMOG).
Macedonia[a]	Tension between Macedonians and Albanians.	(Previously: UNPREDEP).
Myanmar	Democratic opposition to government, ethnic tensions.	
Nigeria	Opposition to a repressive government; ethnic and sectarian violence.	
Pakistan	Government corruption; political violence; sectarian violence.	
Peru	Leftist insurgency led by Sendero Luminoso and MRTA.	
Philippines[a]	Muslim separatist movement in Mindanao.	
Rwanda[a]	Reprisals against Hutus following genocidal attacks on Tutsis.	ICTR.
Russia[a]	Russian response to Chechen secessionist movement.	
Serbia-Kosovo[a]	Tension between Serbs and Kosovar Albanians following Serb "ethnic cleansing" and deployment of NATO forces.	Autonomy enforced and implemented by UN, NATO, and OSCE.

Table 3.1—continued

Country	Character of Conflict	International Response
Sierra Leone	Civil war among government, Army rebels, and RUF.	UNOMSIL, ECOMOG.
Somalia[a]	Clan-based factional fighting exacerbated by droughts.	(Previously: UNOSOM I, Restore Hope, UNOSOM II)
Sri Lanka[a]	Civil war between Sinhalese-dominated government and Tamil separatists.	
Sudan	Civil war between Islamic regime and non-Muslim resistance movements in south.	
Tajikistan	Islamic insurgency supported from Afghanistan.	UNMOT; CIS peacekeeping force.
Turkey[a]	Insurgency of Kurdish separatists led by PKK.	

NOTE; Acronyms are defined on p. xxvii.

[a]Predominant underlying cause is national, racial, or ethnic differences.

five are from Africa (Burundi, Liberia, Sierra Leone, Somalia, and Sudan) and two from Eastern Europe (Bosnia-Herzegovina and Croatia).[3]

Another symptom of conflict is hunger. There is no fundamental scarcity of food; indeed, world food production has increased faster than population over the past fifty years. But some populations are at the margin, and therefore conflict can easily plunge them over the edge into starvation. In 1992, clan warfare in Somalia curtailed distribution of food while a drought decreased the total supply. The result was widespread malnutrition and much outright starvation. A decade ago, the World Food Programme (WFP) gave most of its assistance in ways intended to make people more self-reliant, but today about 70 percent of its assistance goes just to feed victims of conflict. About one-third of food delivery goes to sub-Saharan Africa, where the major recipients in order of volume are Ethiopia, the Great Lakes Region (Burundi, Rwanda, Democratic Republic of the Congo, and Tanzania), Mozambique, Angola, and the Sudan. Another third

---

[3]United Nations High Commissioner for Refugees, "UNHCR by Numbers," http://www.unhcr.cn/un&ref/numbers.

goes to Asia, where the major recipients in order of volume are Bangladesh, the Democratic People's Republic of Korea, Indonesia, and India.

The Former Yugoslavia and Sub-Saharan Africa

The demand for future U.S. intervention is particularly likely to be high in the former Yugoslavia and Africa. Conflicts in the former Yugoslavia and sub-Saharan Africa share some basic similarities. In both regions, national and ethnic antagonisms fuel violent, atavistic struggles for domination that produce brutal discrimination, expulsion, and massacres. In both regions, conflicts cause very large, often extremely sudden refugee flows, which tax the abilities of international organizations and NGOs to respond.

But beyond these similarities are stark differences. Inhabitants of the former Yugoslavia live in Europe and are acting out antagonisms rooted in European history. Indeed, the outbreak of conflict in Croatia and Bosnia was directly related to struggles among fascist, communist, and nationalist groups during World War II. In response, the Western European allies are trying to establish the norms of late-twentieth-century European civilization in the former Yugoslavia, especially in Bosnia-Herzegovina and Kosovo. Sub-Saharan Africa is a different story. The customs, languages, and histories of the peoples are less well understood by policymaking elites in the West. Especially since the humiliating failure in Somalia, these elites have less tendency to believe that even protracted operations could lead to lasting peace and economic development. Moreover, many African countries were once European colonies and their leaders are viscerally opposed to outsider pressures, which they believe resemble colonialism.

Because the former Yugoslavia is in Europe, its affairs are of direct geopolitical interest to major European powers, including three members of the Security Council—Britain, France, and Russia. The United States has little direct geopolitical interest but strong indirect interest because of its leadership of NATO. During recent conflicts, these major powers often diverged on policy, but they remained engaged. By contrast, UN Security Council members have little geopolitical interest in sub-Saharan Africa apart from France's continuing involvement with francophone Africa. They have little fear of refugees. As a result, major powers' engagement is more tenuous in Africa. Through the Security Council and bilaterally, they support the efforts of the Economic Community of West African States (ECOWAS) Cease-Fire Monitoring Group (ECOMOG) in Liberia, Sierra Leone, and Guinea-Bissau, but they avoid large-scale commitment of their own forces and resources comparable to Operation Joint Guard and Operation Joint Guardian.

Conflict in the former Yugoslavia is probably winding down, but conflict in sub-Saharan Africa shows no signs of abating and may well become more intense. As one expert noted:

> Africa, particularly sub-Saharan Africa, is falling apart. It is plagued by overpopulation, poverty, illiteracy, starvation, drought, civil and ethnic war, AIDS, government corruption, crime, deforestation, disease, and everywhere you look, refugees. Sub-Saharan Africa is the "third world of the third world."[4]

In the past decade, civil war has been epidemic across the African continent. Wars started or continued in 15 sub-Saharan countries: Angola, Burundi, Central African Republic, Chad, Congo-Brazzaville, Congo-Kinshasa, Ethiopia, Guinea-Bissau, Liberia, Mozambique, Rwanda, Sierra Leone, Somalia, Sudan, and Uganda. In at least three countries—Liberia, Congo, and Somalia—civil society broke down when rival bands based on ethnic groups competed ferociously and interminably for dominance.

Half of the ten largest refugee populations tracked by UNHCR were generated by conflict in Africa. These people fled from Somalia (524,400), Burundi (515,800), Liberia (486,700), Sudan (315,300), and Sierra Leone (328,300).[5] These conflicts devastate local populations, which are already afflicted by malnutrition and disease. Drought, primitive agricultural techniques, and inefficient government have worsened food security in recent decades, making malnutrition common.[6] About 70 percent of all individuals infected with the HIV virus that produces acquired immune deficiency syndrome live in

---

[4]Bernstein (1994), p. 90.

[5]United Nations High Commissioner for Refugees, "UNHCR by Numbers," http://www.unhcr.cn/un&ref/numbers/table5.num.

[6]IFAD (1993).

Africa.[7] In parts of sub-Saharan Africa, one in every four adults is HIV positive, and the infection is still spreading.

UNPREDICTABILITY OF INTERVENTION

Decisions to actually conduct operations, however, are much harder to anticipate than the strong likelihood that conflicts meriting intervention will continue.   When weighing humanitarian intervention, U.S. officials decide on a case-by-case basis, often unpredictably. For geopolitical reasons, the United States is more likely to mount operations close to home, especially in the Caribbean and Central America, or in areas close to its NATO allies, such as the Balkans and perhaps Cyprus, than in more-distant regions. In addition, the United States remains firmly committed to Korean security and would almost certainly respond strongly to a humanitarian crisis resulting from an implosion or fall of the North Korean regime. But due in part to the Somali debacle, U.S. decisionmakers will probably hesitate to become deeply involved in sub-Saharan Africa, despite that region's turmoil and suffering. Yet even here, the prospect of a truly massive conflict—such as a possible repeat of the 1994 Rwanda genocide—might led to U.S. intervention.[8]

It is difficult to predict when the United States and others will undertake complex contingency operations. Although the United States has an interest in the advancement of human rights, it does not, and cannot, act consistently on such a basis and therefore must choose on more particular grounds. For example, it may want to support its allies, protect U.S. citizens abroad, inhibit or reverse flows of refugees, or counter threats to its prosperity. Policymakers differ widely on definitions of these interests and whether they are sufficiently threatened in particular cases.

When the Balkan wars began in 1991, the Bush administration stayed out, believing that the United States had too little interest to justify

---

[7]Bernstein (1994), p. 90; *Medical Mission News* (Winter 1998), p. 4.

[8]Funding for humanitarian crises mirrors the lack of consistency. For example, the aid effort for Kosovo has received far more aid than can be immediately used, while aid efforts in Africa are neglected.

the massive military effort that intervention would require. In following years, the Clinton administration was equally determined to stay out, even though the European allies were badly floundering. But in 1995, the Clinton administration decided to lead and contribute heavily to a complex contingency operation in Bosnia that would last one year. It subsequently decided to prolong that operation indefinitely. Why did the nation decide to act in 1995 rather than five years earlier? The decision resulted from a complex series of events during the final years of the Balkan wars, which drew the United States ever deeper into a contingency it had initially tried to avoid. In the final analysis, U.S. decisionmakers chose to act because its allies could neither ignore the problem nor solve it without American leadership. In retrospect, these propositions may sound obvious, but they were not obvious when the wars started.

If the United States has difficulty deciding when to conduct operations in Eastern Europe, it has even greater difficulty in sub-Saharan Africa. Why did the Bush administration decide in December 1992 to send forces to Somalia? At the end of the Cold War, the United States was aligned with Somalia in opposition to Ethiopia, a client of the Soviet Union (years earlier the alignments had been reversed). After the collapse of the Soviet Union, the United States had almost no geopolitical interest in Somalia and therefore its primary motive was humanitarian, spurred by scenes of horrendous suffering. But in early 1993, the Clinton administration began to pursue a new goal of better governance, striking at the power of factional leaders, especially in the Mogadishu area. Just a few casualties prompted the administration to abandon this effort, in large part because it could not identify interests that would justify the effort.

The Rwandan crisis, which began on April 6, 1994, presents an even more complicated picture. During the first three weeks, extremist Hutus hacked to death with machetes, shot to death, and burned alive almost a million people, throwing tens of thousands of bodies into the rivers.[9] At the time, the United Nations Assistance Mission for Rwanda (UNAMIR) had some 2100 personnel, too few to be decisive. The U.S. government believed there was insufficient

---

[9]United Nations, *The Blue Helmets* (1996), pp. 346–348.

international or domestic support to mount a successful operation and urged the reduction of UNAMIR. The Security Council drew UNAMIR strength down to 270 personnel and gave them the mission of acting as monitors and intermediaries between the parties. The French government subsequently launched its controversial Operation Turquoise in southwestern Rwanda but soon relinquished control to the Tutsi-led Rwandese Patriotic Front (RPF). During June and the first week of July, the RPF won decisive victories in Rwanda. Beginning on July 3, 1994, some three million panic-stricken Hutus, including most of the perpetrators of the genocide, fled to neighboring countries—one of the most sudden and massive refugee flows in history. Refugees who had fled to the neighboring Congo (then Zaire) lacked water, food, and sanitary facilities. Under these conditions, they began dying rapidly of cholera. On July 17, the Clinton administration decided to provide humanitarian assistance to these refugees in Operation Support Hope. Thus, the United States decided not to forcefully oppose Hutu genocide against Tutsis but subsequently mounted an operation to save Hutus, including an exile regime that dominated the refugee camps while planning a return to Rwanda. These U.S. decisions were understandable responses to specific emergencies, but no one would have predicted them.

The type of mission conducted, as well as the decision to intervene, is also difficult to predict. The United States supported a massive relief and political effort in response to the Somalia famine and civil strife but engaged in only modest relief activities after the bloody and brutal Rwanda genocide. Moreover, goals regularly change during an operation as new information emerges or as political leaders realize opportunities to resolve the conflict in the long term or recognize unanticipated constraints on military effectiveness.

DRIVERS OF INTERVENTION

In deciding to conduct operations, U.S. decisionmakers are driven as much by domestic opinion and allied concerns as humanitarian motives. Both are varied and unpredictable. Fundamental decisions—when and where to conduct operations, what goals to set, how many resources to invest, when to terminate operations—are made with domestic and allied audiences in mind.

Media coverage and grassroots efforts by NGOs can put great pressure on the U.S. government to act.[10] In Somalia, Kosovo, and other crises, the highly publicized suffering prompted popular support for intervention. Moved by media reports of suffering abroad, local civic and church groups often collect goods—some of little immediate value in the crisis—and press politicians to ensure they are shipped promptly.

Media coverage, however, is uneven. Kosovo, for example, received tremendous attention while bloodier conflicts such as Sudan or the Ethiopia-Eritrea war received almost none. Often, other news events relegate a humanitarian crisis to obscurity. In many crises, the initial suffering receives tremendous attention, but the media coverage wanes in subsequent months.

Refugee flows and associated allied concerns are another major driver of intervention. Particularly in Europe, U.S. allies push for action because they fear massive flows of refugees into their own countries and the spread of violence. The United States also responded to Turkey's concerns about Kurdish refugees after the Iraqi civil strife in 1991, prompting Operation Provide Comfort.

LIMITED WARNING TIME

Crises may occur with little warning for policymakers or military leaders because intelligence on crisis regions, particularly in sub-Saharan Africa, is often limited. Limited U.S. and allied interests in the region result in few diplomatic or intelligence assets being devoted to events there. Moreover, the civil war or bad government that caused the crisis often leads the United States and its allies to reduce their already-limited official presence for safety reasons. Thus, the U.S. government had withdrawn its official presence from Somalia in the months preceding its major intervention there.

At times, the United States can anticipate *where* complex emergencies are likely to occur, but seldom *when* they will occur. In 1990, the United States saw that Yugoslavia was about to break apart and believed that violent ethnic conflicts would ensue. Indeed, the U.S.

---

[10]Menkhaus (1998), p. 55.

government based its policy on that expectation. Tension was clearly apparent in Rwanda long before the genocide and the United Nations was working, albeit ineffectually, to alleviate tension and implement a peace plan.

But events often take capricious courses. They may postpone crises that appear imminent and precipitate crises that appear remote. Policymakers had little warning that civil war would break out and rapidly engulf Bosnia in the spring of 1992. They did not anticipate that Kosovar Albanian terrorism and Serb repression would spiral out of control in early 1999. They had no warning that the presidents of Rwanda and Burundi would both die in an airplane crash in April 1994, precipitating a planned genocide.

Even if U.S. policymakers could anticipate where and when a complex emergency would occur, they still might not prepare an adequate response. Complex contingency operations are usually international and multinational, involving a large number of actors outside the U.S. government who are unlikely to agree on a course of action until confronted by an actual crisis. Even within the U.S. government there can be strong resistance to advance planning, unless demanded at the highest levels. Many policymakers are too absorbed by current crises to plan for future crises, and they may simply prefer to keep their options open.

PREFERENCE FOR MULTILATERALISM

Seldom, if ever, will the United States mount a complex contingency operation unilaterally or bilaterally. Even in cases where refugees are not of immediate concern to U.S. allies, they often share U.S. humanitarian objectives because the same media concerns that drive U.S. decisionmakers also affect other Western publics. Almost always the operation will be international, involving a colorful cast of actors. These actors include host countries, whose governance may be a central issue. They almost invariably include other major powers whose interests may be involved, especially other permanent members of the Security Council. They may include regional actors whose support will be critical, as Albania currently supports Operation Joint Guardian. Donor countries, principally the West European countries and Japan, that finance key activities such as refugee return, social programs, and reconstruction also play a vital role. Troop contributors, some of whom are minor powers with extensive experience in peace operations, such as Austria, Canada, Finland, India, Norway, Pakistan, and Sweden, are important partners. Actors almost always include organs of the United Nations and its family of organizations, the World Bank Group, the International Monetary Fund, and regional security organizations such as ECOWAS, NATO, the Organization of African Unity, the Organization of American States, and the Organization for Security and Cooperation in Europe (OSCE). These are all important players in the overall mission.

Chapter Four

MILITARY TASKS IN COMPLEX CONTINGENCIES

The military will normally play a supporting role during crises, helping relief agencies provide assistance rather than taking the lead. Relief agencies are often better able to carry out assistance tasks due to past experience. Moreover, policymakers may restrict the length or extent of military participation in these operations. As a matter of principle, both military and civilian officials prefer that the military should accomplish tasks unrelated to its core mission only on an exceptional basis, i.e., when no civilian agency can do the job quickly enough or well enough under the circumstances. But exceptions can be the rule during the first phases of complex contingency operations, while the situation remains unstable and civilian agencies are not yet fully able to carry out their responsibilities.

Although the range of possible military tasks in complex contingency operations is vast, they usually fall into five general categories of activities: (1) providing humanitarian assistance; (2) protecting humanitarian assistance; (3) assisting refugees and displaced persons; (4) enforcing a peace agreement; and (5) restoring order.

Military responsibilities regularly exceed the initial taskings.[1] In Provide Comfort, military forces were initially tasked to provide water and food to the Kurds; soon they were called on to transport and secure returning Kurdish refugees. Forces also had to ensure the withdrawal of Iraqi military forces. Thus, if preparations are limited

---

[1]Siegel (1995), p. 2.

to only the initial tasking, military forces may not be prepared for subsequent, more ambitious objectives.[2]

PROVIDE HUMANITARIAN ASSISTANCE

Civilian agencies, including international organizations (IOs), the Red Cross, and nongovernment organizations, provide most humanitarian assistance with little or no help from the military. But sudden natural catastrophes and human conflict can temporarily overwhelm civilian agencies and require the military to help in several ways.

Conduct Airlift and Sealift

One of the most important forms of assistance is transporting relief supplies to crisis regions. In general, ground transportation is done with local NGO and IO assets. The military can help coordinate the logistics effort, particularly in the early days of a crisis when operations are chaotic. The U.S. military also conducts hundreds of relief flights annually, including dedicated missions and space-available flights under the Denton Program.[3] Moreover, Air Mobility Command frequently mounts larger operations, which may be purely humanitarian like Operation Support Hope in Rwanda or part of a larger operation with humanitarian aspects like Operation Joint Forge in Bosnia.[4] The U.S. military, usually in coordination with host countries or allies, may also take wider responsibility for ensuring adequate airflow. It may establish air traffic control, provide navigation aids, improve airports and related facilities, and offload arriving planes. During Operation Support Hope, for example, the United States improved airports at Entebbe and Goma. It also tried to exercise overall air traffic control through an operations center established at UNHCR headquarters in Geneva. On rare occasions, the U.S. military may even airdrop humanitarian supplies. For

---

[2]Siegel (1995), pp. 31, 52.

[3]The Denton Program allows the Department of Defense to provide space-available transportation of humanitarian cargo at little or no cost to the donor.

[4]In fiscal year 1998, for example, the USAF provided relief supplies to Bosnia, food relief to Germany, peacekeepers to Macedonia, earthquake and flood relief to China, and medical supplies to Albania and Russia, and helped fight forest fires in Indonesia and Ecuador, among other operations.

example, the United States and two of its NATO allies airdropped supplies to Muslim-held areas in Bosnia, which were isolated by Bosnian Serb forces.

The United States normally provides sealift through the Military Sealift Command just to the military component of a contingency operation, such as NATO forces in Bosnia during Operation Joint Endeavor and Operation Joint Guard. Exceptionally, it may employ military sealift for humanitarian purposes, as in Somalia. The United States Navy and Coast Guard may also improve and operate seaports, as in Mogadishu during Operation Restore Hope and Port-au-Prince during Operation Uphold Democracy.

Provide Specialized Functions

To support its own operations, the military contains a wide variety of specialized functions, making it a self-sufficient community. During a complex contingency, the military may have to perform some of these functions for civilian populations until other agencies can meet local needs. The military may make communications available to host countries, IOs, and NGOs, especially in countries whose telecommunications systems are inadequate or devastated by war. In sub-Saharan Africa, for example, IOs and NGOs may depend upon the military for reliable communications long into an operation. The military may provide medical care, including inoculation, triage, first aid, hospitalization, and medical evacuation. It may promote public health through such means as sanitation and water purification. It may provide means to disseminate public information, perform civil engineering tasks, fight fires, or do practically anything else within its wide-ranging capabilities until civilian agencies are able to perform these functions adequately.

Provide Vital Supplies

Especially during the initial phases of an operation, the military may provide support directly to an afflicted population or to relief agency personnel. This support may include distributing humanitarian daily rations, potable water, building materials, and other supplies. It may also involve providing building materials for shelters.

PROTECT HUMANITARIAN ASSISTANCE[5]

Providing humanitarian aid is not enough to end a crisis when the suffering is caused by human agency—the relief supplies must also be protected. NGOs and UN agencies are often not able to protect relief with their own assets, and at times their efforts to hire private security make the overall environment less secure.

Belligerents often try to obstruct, divert, and even pillage humanitarian aid. They may want to deny aid to their enemies, reward their own supporters, or simply enrich themselves. Particularly in Somalia and Liberia, factional leaders plundered humanitarian aid, demanded a percentage for allowing aid to pass, and compelled aid organizations to hire their supporters as guards, in effect extorting bribes.[6] Bosnian Serbs often prevented UN agencies from reaching isolated Muslim communities or insisted on receiving comparable aid themselves, regardless of need. To secure humanitarian assistance, the military may also have to provide security for airports and seaports, where supplies initially arrive; for internal distribution, including warehouses, convoy routes, and distribution points; for personnel from IOs and NGOs that are basic providers of humanitarian aid; and possibly for "safe areas."

Secure Airports and Seaports

The military may have to secure airports and seaports where humanitarian aid arrives. For example, in March 1992, a relief ship chartered by the United Nations tried to land at Mogadishu, but was

---

[5]Joint doctrine distinguishes four "peace enforcement" missions: enforcement of sanctions, protection of humanitarian assistance, operations to restore order, and forcible separation of belligerent parties. Joint Chiefs of Staff (1999), pp. III-4 to III-6. But protection of humanitarian assistance may not be related to peace. Military forces could protect humanitarian assistance for a time while conflict continued or remained ready to break out again, as indeed happened in Somalia. Operations to restore order go well beyond peace, meaning in this case the cessation of armed conflict.

[6]The same drought had devastated neighboring Ethiopia and northern Kenya, but only Somalia suffered massive casualties, because chaos made it impossible to deliver relief. In Somalia, a key military task was to secure ports, airfields, and lines of communication to the interior in order to help the delivery of relief. Dworken (1995); Dworken (1996); Natsios, "Humanitarian Relief Intervention in Somalia" (1997), p. 79; Kennedy (1997), p. 100.

driven away by fire. In October of the same year, the United States suspended its airlift after one of its aircraft was hit in Mogadishu. To prevent such interference, U.S. forces secured the seaport and airport during Operation Restore Hope.

Local combatants may also target relief aircraft, compelling the military to protect aircraft in the local air space. For example, Bosnian Serb forces initially tried to prevent relief flights from landing at Sarajevo. NATO eventually succeeded in enforcing an exclusion zone around the city where the belligerents were prohibited from employing heavy weapons, although aircraft remained vulnerable to Serb SAMs.[7]

Secure Distribution of Relief

Internal distribution of relief supplies can present greater difficulties than does the initial delivery. In a lawless environment, any armed group may attempt to plunder or divert relief supplies.[8] In Somalia, for example, militias and bandits systematically plundered relief supplies and sold them to local merchants who offered them for sale on the open market.[9] In Somalia, the Sudan, and elsewhere, several large relief organizations found that over 80 percent of food supplies were lost due to theft or misappropriation.

IOs and NGOs typically take a much different attitude toward banditry and graft than does the military, creating a potential source of friction. In Somalia, relief agencies often hired local fighters to act as guards, whom UN forces then attempted to disarm.[10] As one interlocutor noted, major NGOs often "don't want to shoot people for taking the food that they brought." On this principle, NGOs would avoid confrontation that the military would routinely accept.[11] UNHCR makes this comment:

---

[7]On September 3, 1992, Serb forces used a missile to down an Italian G-222 cargo aircraft.

[8]Authors' interviews.

[9]Natsios, *U.S. Foreign Policy* (1997), p. 83.

[10]Kennedy (1997), p. 111.

[11]Authors' interview with NGO official.

> Because of the need to negotiate with armed groups for access to displaced people and other conflict-affected populations, aid agencies often implicitly accept that a proportion of their relief will go to the very groups which are waging the war.[12]

But the military cannot adopt such a permissive attitude. Indeed, it would negate the very purpose of military forces if they allowed militias and bandits to plunder with impunity.

Protect IO and NGO Personnel

IO and NGO personnel normally rely on their neutrality and impartiality to protect them, but in some situations they may have to rely on military forces for security. Belligerents may try to disrupt the aid flow by attacking the aid providers or by intimidating them. In Angola, Burundi, Chechnya, Rwanda, Sierra Leone, Sudan, and other countries, belligerents have attacked and murdered personnel from IOs, the Red Cross, and NGOs.[13] Since January 1992, 184 UN civilian aid workers have been killed. Belligerents may target relief efforts because they oppose a peace operation that is simultaneously in progress. In February 1994, for example, a Somali militia leader bombed the headquarters of World Vision in Baidoa because he opposed the United Nations–led peace operation, even though it was unrelated to World Vision's operations.[14] In Kosovo, the Serb government arrested several CARE workers on espionage charges.

The military may need to secure NGO personnel, guarding their quarters, escorting them on the road, and dealing with "warlords" who try to intimidate them.[15] Securing NGO personnel can be exceptionally difficult because NGOs usually have to disperse their workers in order to accomplish their missions. For example, NGOs had 585 offices, residences, feeding centers, clinics, and other facilities scattered throughout Mogadishu during Operation Restore

---

[12]UNHCR (1997), p. 45.

[13]UNHCR (1997), pp. 48 and 132.

[14]Natsios (1995), p. 72.

[15]Kennedy (1997), p. 100.

Hope. NGOs refused to consolidate their activities because they wanted to maintain close contact with the local population.[16]

The situation becomes more complicated when IOs and NGOs are compelled to hire security guards who are themselves involved in banditry and lawlessness. In Somalia, for example, during the height of the anarchy, IOs and NGOs could not operate without armed guards recruited from clan militias. At peak, the International Committee of the Red Cross (ICRC) hired thousands of armed guards in Somalia.[17] In an attempt to impose order, however, the military prohibited open display of weapons. As a result, it frequently disarmed security guards, causing friction with the IOs and NGOs who had hired them.[18] After U.S. troops left Somalia in March 1994, the situation deteriorated to the point that clan leaders demanded vehicles and supplies before they would allow UNOSOM II to depart peacefully. To secure the departure of UNOSOM II, the United States conducted Operation United Shield, which involved U.S. Marines, Italian troops, and special operations forces.

Establish Safe Areas

One technique for securing relief is to declare safe areas where the population is protected from the effects of conflict. But this technique may require more force than outside powers are willing to apply. During Operation Provide Comfort in 1991, the United States protected areas in Northern Iraq where Kurds were safe from Iraqi forces. While temporarily helpful in returning refugees to their homes, these areas could not be defended indefinitely without large forces—in 1996, Iraqi forces overran part of the protected area. The "safe areas" declared by the United Nations in Bosnia were hardly secure, and Bosnian Serb forces overran two of them.[19] The former chief of UNHCR's Bosnia operation noted that these safe areas were:

---

[16]Dworken (1995), p. 17; Natsios, "Humanitarian Intervention" (1997), p. 92.

[17]Natsios, "Humanitarian Intervention" (1997), p. 84.

[18]Dworken (1995); Kennedy (1997), p. 111.

[19]Security Council Resolution 824, passed on May 6 1994, declared six safe areas: Bihac, Gorazde, Sarajevo, Srebrenica, Tuzla, and Zepa. Of these safe areas, Bihac was partially invaded; Gorazde was partially invaded; Sarajevo was often bombarded; Srebrenica was overrun; and Zepa was overrun.

. . . surrounded by enemy forces, without basic shelter, medical assistance or infrastructure, isolated and living under sporadic shelling or sniper fire, these areas are becoming more and more like detention centres, administered by the UN and assisted by the UNHCR.[20]

ASSIST REFUGEES AND INTERNALLY DISPLACED PERSONS

Refugees are persons who have fled across international borders, due either to conflict or to fear of persecution. Internally displaced persons (IDPs) have fled for similar reasons, but within the territory of a recognized state. In many instances, belligerents have deliberately caused flows of refugees to exact revenge or to permanently conquer territory. Such refugee flows occurred as a result of conflicts in Bosnia, Burundi, Croatia, Georgia, Serbia (Kosovo), and Rwanda.

Military forces may help establish refugee camps, secure these camps and keep order within them, and support return or resettlement of refugees. These tasks are similar to those required for providing and protecting humanitarian assistance, but they have their own distinct requirements. NGOs and IOs often take a leading role in assisting refugees, although military assistance may be vital to their efforts.

Construct and Maintain Camps

When refugees arrive suddenly in large numbers they may overwhelm local resources, requiring the construction of purpose-built refugee camps. UNHCR usually takes the lead in constructing such facilities but may require military assistance. During the Kosovo crisis in early 1999, for example, the United States and several other militaries built refugee camps in Albania to house tens of thousands of refugees. As part of Operation Shining Hope, the U.S. military constructed Camp Hope with a capacity for 20,000 refugees, under management by CARE.

---

[20]UNHCR (1997), p. 136.

Protect Refugees and IDPs

Refugees and IDPs may require special protection. They may be attacked by belligerents, harassed by host countries, subjected to involuntary conscription, exploited, abused, and plundered. Groups that have violent agendas may dominate refugee camps. Normally, IOs and NGOs administer camps without assistance from the military, but in some cases, the military may need to provide security. For example, the U.S. military provided security in camps for Cubans and Haitians fleeing their respective countries.

Local people may also resent the refugee population and therefore harass refugees or support their enemies. Medécins Sans Frontières (MSF), for example, noted in November 1996 that Zairians resented Rwanda refugees because of "the living daily parody that the refugees in camps have a far better quality of life."[21] As a result, these locals often supported anti-refugee forces from Rwanda.

Refugee camps can become new sources of violence and instability. Criminality may become epidemic in the camps, including theft, rape, and murder, as occurred in Somalia and Zaire.[22] Warlords may recruit among refugees, and camps can even become military bases, with their humanitarian status being used to guarantee a safe haven. In Zaire, for example, the Rwandan government regularly battled Forces Armees Rwandaises (FAR) marauders based in UN-run refugee camps.[23] In Somalia, many food distribution centers were in the area controlled by Mohammed Aideed. As a result, people from other areas moved to Aideed's area of control, strengthening his power.[24]

Sometimes, refugee camps cause new flows or make it less likely that refugees will return home. When individuals have lost their homes and livelihoods, a refugee camp may be their best short-term option. But it can be easier to keep receiving free food and shelter than to start over again in an insecure and demanding environment. Some

---

[21] As quoted in UNHCR (1997), p. 73.

[22] Natsios, "Humanitarian Intervention" (1997), p. 80.

[23] Boutroue (1998), pp. 4–5.

[24] Natsios, "Humanitarian Intervention" (1997), p. 88.

people may even be attracted to refugee camps that offer better conditions than in their own towns and villages.

An extensive information campaign may be required to inhibit new flows of refugees and IDPs. Exaggerated reports, rumors, and misperceptions may cause people to flee from nonexistent threats or to seek help that will not materialize. For example, Hutu extremist propaganda convinced many Hutus that the new Rwandan government would butcher them, causing many to flee the country. Counteracting such propaganda was an important mission for intervening agencies during Operation Support Hope.[25]

Support Return and Resettlement[26]

Eventually, refugees and IDPs must either return home or find new homes elsewhere. Their return is often essential to complete resolution of conflict. In Bosnia, Kosovo, and Rwanda, return of refugees was a major concern to policymakers negotiating a political settlement.

Refugees will not return if their security is threatened. In northern Iraq, Kurds refused to go home until they were sure that Baathist forces had left. In Bosnia, most refugees still refuse to return if their former homes lie in a region dominated by another ethnic group. Military forces can ensure safe return, but they seldom can stay long enough and in sufficient strength to protect the returned refugees. As a result, there have been relatively few minority returns in Bosnia despite a concerted international effort.

Members of a refugee group may also obstruct or oppose return. In Rwanda, FAR members compelled Hutu refugees to remain in camps in order to shield themselves from Tutsi forces and to escape punishment for genocide. UNHCR did not control these camps and

---

[25]U.S. European Command, *Operation Support Hope*, p. 10.

[26]Joint doctrine subsumes resettlement of civilian refugees and displaced persons under the mission of forcibly separating belligerents. Joint Chiefs of Staff (1999), p. III-6. But resettlement might not separate belligerents. In Bosnia, for example, minority returns *mix* ethnic groups that were belligerents, rather than separating them.

therefore could not separate those guilty of genocide from those who had fled for other reasons.[27]

ENFORCE PEACE

Enforcing peace may include applying sanctions, separating formerly belligerent forces, and disarming formerly warring factions. Enforcement often centers on peace agreements concluded by the parties to conflicts. Such agreements normally include military provisions intended to prevent or inhibit fresh outbreaks of conflict. The military may perform many of these tasks in association with the United Nations.[28]

Apply Sanctions

The military may enforce sanctions against belligerents or parties to a conflict, usually under resolutions of the Security Council. Sanctions may include restrictions on the movement of civilian goods, on the introduction of arms and military supplies, and on traffic generally.

Since August 17, 1990, the United States and its allies have enforced an economic embargo against Iraq through maritime interception operations under authority of several Security Council resolutions. At peak, ships of 14 countries helped enforce this sanction.[29] In

---

[27]Gourevitch (1998), pp. 166–167.

[28]Operations in Somalia were based in part on agreements concluded among the factions in Addis Ababa on January 15 and March 27, 1993. The entry of U.S. forces into Haiti was based on an agreement concluded between U.S. emissaries and the regime in Port-au-Prince on September 18, 1994. The primary mission of NATO forces in Bosnia was set forth in Annex 1A of the Dayton Agreement, initialed on November 21, 1995. The mission of NATO forces in Kosovo was set forth in a paper presented to the Belgrade regime on June 2, 1999, an implementing Military Technical Agreement signed on June 3, and Security Council Resolution 1244 adopted on June 10.

[29]An increasingly common sanction is the creation and enforcement of a "no-fly" zone. In April 1991, the United States and several allies initiated Operation Provide Comfort, initially to ensure compliance with Security Council Resolution 688, which demanded that Iraq cease repressing Kurds in northern Iraq. During this operation, the United States and several allies deployed forces in northern Iraq to protect Kurdish refugees and to assure their safe return. This operation was succeeded by Operation Northern Watch, which enforced a no-fly zone north of the 36th parallel. British, Turkish, and U.S. forces currently enforce this sanction through air operations. In

September 1991, the Security Council adopted Resolution 713, proclaiming a general embargo on all deliveries of weapons and equipment to Yugoslavia. This resolution was aimed at the Federal Republic of Yugoslavia (essentially Serbia), which was then attacking newly independent Croatia. NATO enforced this sanction in the Adriatic through Operation Maritime Monitor and subsequently through Operation Maritime Guard.[30]

Separate Belligerent Forces

The military may enforce provisions of peace agreements to separate the forces of formerly belligerent parties, often through buffer zones where the parties are not allowed to deploy forces. The United Nations traditionally monitors, but does not enforce, buffer zones. Examples include the zone of separation between Israel and Syria on the Golan Heights, monitored since June 1974; the buffer zone between Greek and Turkish Cypriots on Cyprus, monitored since August 1974; and the demilitarized zone between Iraq and Kuwait, monitored since May 1991. In contrast to the United Nations, NATO has undertaken to enforce buffer zones. Examples include the zone of separation between the Croat-Muslim Federation and the Serb Republic in Bosnia, enforced since January 1996; and air and ground safety zones between Serbia and its Kosovo province, enforced since June 1999.

August 1992, the United States and several of its allies began enforcing a no-fly zone in Iraq south of the 32nd parallel (Operation Southern Watch). Southern Watch was initially under authority of Security Council Resolution 688, adopted in the preceding year. In October 1994, Saddam Hussein deployed forces in Southern Iraq in a manner that threatened Kuwait. The United States came to Kuwait's defense in Operation Vigilant Warrior. Thereafter, the Security Council adopted Resolution 949, which prohibited Iraq from deploying units south of the 32nd parallel. The United States and several allies currently enforce these sanctions through air operations.

[30]After a three-sided conflict broke out in Bosnia, this sanction worked in favor of the Bosnian Serbs, who had inherited equipment from the Yugoslav Army, and against the newly independent Bosnia. Under these changed circumstances, the United States regarded the embargo as highly immoral. It lobbied strenuously to allow Bosnia to import arms, but its European allies, who had ground forces deployed in Bosnia, refused to lift the sanction. In August 1994, the U.S. Senate voted to stop funding enforcement of the embargo if Bosnian Serbs would not agree to a peace plan. On November 10, the United States announced its unilateral decision to stop enforcing the embargo through naval operations. However, the majority of arms shipments reached Bosnia either by air or through Croatia, whose forces extracted a toll on all arms and on goods going through its territory.

Disarm Belligerent Forces

The military may enforce provisions of peace agreements to disarm, demobilize, and demilitarize forces of formerly belligerent parties. In Somalia, U.S. forces confiscated certain types of weapons, especially "technicals" (jeeps and trucks with heavy machine guns mounted on them), and began licensing all weapons. In Haiti, U.S. forces confiscated unauthorized weapons and conducted a program to purchase weapons from civilians. In Bosnia, NATO forces helped enforce arms limitations under the Dayton Agreement. In Kosovo, they disarmed the Kosovo Liberation Army (KLA) under an agreement concluded with its leadership.

RESTORE ORDER

The military is often given vague instructions, such as restoring order or establishing a secure environment. Often this is done in conjunction with the United Nations, but relief agencies seldom play a major role. For good reasons, the military would often like to avoid such a mission. It represents an unlimited, open-ended responsibility, which may be difficult to relinquish safely. It entails tasks appropriate for indigenous police, not foreign troops. But in the initial phases of a complex operation, there may be no alternative to military forces. Even in subsequent phases, military forces may have to accomplish tasks that exceed local capabilities or provide an ultimate guarantee of stability.[31]

Halt Violence

Especially during the initial phase of an operation, military forces cannot escape responsibility for maintaining public order. On

---

[31]Assuring a secure environment may have very different implications, depending upon the situation. In parts of Somalia, government had collapsed entirely, resulting in near-anarchy and rule by armed militias. In Haiti, much of the government, especially police and judiciary, was either hopelessly corrupt or compromised by association with the Duvalier and Cédras regimes. In Bosnia, working governments were already in place, but they were protective of their own ethnic groups and unwilling to cooperate with each other. In Kosovo, NATO forces initially encountered de facto control by former KLA members or no government at all after Serb officials had departed.

December 20-21 1989, U.S. forces quickly seized Panama against sporadic, occasionally fierce resistance. But following the invasion, looting and violence broke out in Panama City, requiring military police to intervene.[32]   When U.S. forces arrived in Haiti on September 19, 1994, the Chairman of the Joint Chiefs of Staff announced that the task of keeping law and order would be the responsibility of the Haitian police and military. But after Haitian police attacked demonstrators, President Clinton changed the rules of engagement, granting tactical commanders discretion to curb violence. On September 24, a U.S. Marine patrol killed the occupants of the main police station in Cap-Haïtien after they brandished weapons. Thereafter, Haitian civilians drove the police away and ransacked the police stations, leaving an authority vacuum.

In Bosnia, NATO forces often intervened to maintain public order, especially when violence broke out between ethnic groups. A special concern was Brcko, a town that had been predominately Muslim prior to the war. Serbs had seized the town and supplanted its Muslim population because Brcko lay at the slender throat of a corridor connecting the western and eastern halves of their territory. Brcko was placed under international control pending arbitration— which eventually went in favor of the Croat-Bosniac Federation. One U.S. battalion maintained a constant presence in Brcko in a largely successful attempt to prevent ethnic violence.

Reinstate Civil Authority

Military forces may have to provide security for public officials and during elections and parliamentary sessions. In Haiti, U.S. forces provided security for Jean-Bertrand Aristide, the democratically elected President who had been living in exile since a military coup, and for government buildings in Port-au-Prince. Aristide's assassination, very possible in the prevailing environment, would have deprived democratic forces of a charismatic leader and possibly caused a spiral of violence. In Bosnia, NATO forces provided security for moderate Serb leaders in Banja Luka who opposed hard-liners in Pale. For example, on October 8, 1997, NATO troops and attack heli-

---

[32]Taw (1996), p. 15.

copters intervened to prevent hostilities between police loyal to the newly elected President Biljana Plavsic and those loyal to Radovan Karadzic, the wartime leader indicted for war crimes. Within a few months, the Pale faction lost power and Karadzic became a fugitive.

Military forces frequently have to provide security for electoral activity, including registration, political campaigning, and casting of ballots. The major accomplishment of UN forces in Cambodia was a relatively peaceful general election conducted in May 1993. In June 1996, 3000 NATO troops helped assure that municipal elections were conducted peacefully in Mostar, despite a bitter division between Croats and Muslims.

Assist Police Forces

Military forces may help police forces to recover their strength and reassert their authority. Military forces, especially military police, may conduct joint patrols with indigenous police. They may help to equip and to some extent train newly established police. In addition, military forces may have to respond quickly during crises, such as widespread looting and riots, which temporarily overwhelm indigenous police.

After the U.S. invasion of Panama, Southern Command launched Operation Promote Liberty through a Civil-Military Operations Task Force controlled by the U.S. Charge d'Affaires.[33] The United States quickly reconstituted the disbanded Panamanian National Police, and U.S. military police provided training until Congress reaffirmed then-current legislation that prohibited the U.S. military from training foreign police.[34] Thereafter, the International Criminal Investigative Training Assistance Program (ICITAP) of the Department of Justice assumed responsibility.

During Operation Joint Endeavor and Operation Joint Guard in Bosnia, the various ethnic communities already had well-established police forces, but these were prone to favor their own ethnic groups

---

[33]Oakley et al. (1998), p. 51.

[34]Section 660 of the Foreign Assistance Act of 1961 prohibited U.S. military forces from training, equipping, or advising foreign police. It was repealed in 1995.

and could be overwhelmed easily. The International Police Task Force had a mandate to monitor and advise police but not to exercise police powers. As a result, there was a wide gap between indigenous police forces, which would normally be on hand but had limited capabilities, and NATO forces, which had great capabilities but could respond only in exceptional circumstances. To fill this gap, NATO ambassadors agreed to establish a Multinational Specialized Unit of paramilitary forces, including Italian Carabinieri. During the initial phase of Operation Joint Guardian in Kosovo, practically no police forces existed after the flight of Serbian policemen from the province. As a result, NATO ground forces found themselves in the unwelcome and onerous role of policing Kosovo until indigenous police could be established.

Restore Civil Infrastructure

Donor countries, IOs, and NGOs often assume responsibility to improve infrastructure and generally to revive economic life after conflict. As a general principle, civilian agencies are preferable to military forces for these responsibilities because they are less expensive and employ more local labor. But on an exceptional basis, the military may repair damaged infrastructure or construct new facilities. In the initial phase of an operation, the military often has to repair transportation infrastructure such as airports, seaports, roads, and bridges just to facilitate its own operations. Such improvements also help revive the civilian economy.

In Somalia, for example, U.S. and UN forces found themselves in an already poor country that had been systematically looted and exploited by opposing factions. To support their own operations and delivery of humanitarian aid, they had to repair infrastructure, including over 1,000 kilometers of road and the Sean Devereux Bridge near Kismayo. During operations in Haiti, the U.S. military restored electrical power in Port-au-Prince, improved seaports, repaired roads, and renovated public buildings. In Bosnia, NATO militaries repaired and expanded airports, repaired and replaced bridges, and assisted in a wide range of public construction projects.

As the brief description presented in this chapter suggests, complex contingency operations cover a vast range of activities and missions. Many, if not most, of these operations involve military support to

IOs, NGOs, or host governments. Such activities require establishing a partnership and anticipating potential problems in the relationship, which are discussed in the following chapter.

Chapter Five

COMMON CONSTRAINTS ON OPERATIONS

During complex contingency operations, the military experiences constraints and problems that would not usually be present to the same extent during war or in response to relief in a simple natural disaster. As discussed in Parts Two and Three, these constraints also have implications for coordination with relief partners and allies, leading to problems with advanced planning, difficulties explaining the importance of force protection, and challenges to relief agency impartiality. This chapter describes constraints and problems common to complex contingency operations.

WEAK RESOLVE AT HOME

In most complex contingency operations, the United States and its allies have few if any vital or important geopolitical interests at stake.[1] Absent important interests, the United States and its allies tend to have weak resolve. Such limited resolve is particularly likely when intervening in sub-Saharan Africa, where the United States and its allies have almost no interests at stake. Weak resolve can produce the following problems and constraints:

• Aid is often a poor substitute for political action.

• Requirements are often skewed toward political rather than humanitarian objectives.

---

[1]Gow (1997), pp. 299–300.

- Casualty sensitivity limits the types of missions and resource allocation.

- Restrictive rules of engagement (ROE) are common.

Humanitarian Aid as a Substitute

Confronted with situations that demand action yet reluctant to become deeply involved, major powers offer humanitarian aid as a substitute for a more comprehensive strategy for solving a region's problems. As a UNHCR report argued, "Unfortunately, states have tended to use humanitarian action as a substitute for political action rather than as a complement to it."[2] Thus, when the Bush administration announced its decision to mount an operation in Somalia in December 1992, it cited humanitarian motives engendered by the horrendous suffering there. However, it proposed to secure relief, not to address the underlying cause of suffering—interminable fighting among several clans. Similarly, instead of trying to enforce peace on the belligerents in Bosnia, European powers stressed UNPROFOR's role in securing humanitarian aid, because these powers feared the consequences of deeper involvement.

Substituting aid for political action is often futile. Indeed, trying to secure humanitarian aid during conflict is a risky task and may even be counterproductive.[3] Belligerents quite naturally regard relief supplies as sources of power. Food, medicine, and other supplies are highly valuable in war-torn regions, and those who control them can increase their power. Warlords want to secure supplies for themselves and deny them to their enemies. Belligerents may have no qualms about looting humanitarian relief agencies or extorting supplies as the price for allowing these agencies to operate. Sometimes, humanitarian aid can even increase suffering by supplying the forces of belligerents. During the Somali civil war, the provision of aid bolstered clan-based militias, through looting and through payment for protection. In Liberia, warlords deliberately impoverished and displaced local communities to attract aid for the

---

[2]UNHCR (1997), p. 44.

[3]Barber (1997).

victims.[4] During the Bosnian conflict, humanitarian aid helped all parties prolong the struggle.[5]

The most egregious example of disastrous good intentions was the international support to Hutu refugees in Zaire during Operation Support Hope. Among those refugees were thousands of militiamen, soldiers, and government officials who had recently committed acts of genocide against the Tutsis in Rwanda. Thanks to international aid and the accommodation of Zaire, they found a haven in Zaire from which to organize, arm, and launch raids into Rwanda.[6] These activities came to an end when Tutsi forces invaded Zaire, precipitating a civil war that toppled the decrepit Mobutu regime.

Skewed Requirements

Political requirements also may lead to pressures on the relief effort that planners should anticipate. Policymakers seeking to sustain support for an operation may need to show immediate results. Success may be measured by how impressive the operation appears on television rather than humanitarian measures of effectiveness, such as the number of refugees returned to their homes and drops in morbidity rates.[7] Host country officials may prefer that foodstuffs and other visible evidence of a relief effort arrive before forklifts, K-loaders, and other items that would increase overall through-put and perhaps save more lives. Often, U.S. and allied governments seek immediate credit for alleviating a humanitarian disaster in order to reap political rewards.

Political concerns are present at all levels throughout the system. Local civic groups may press their representatives to force the military to transport their donations, even if this does little to help the victims. Local politicians often see humanitarian crises in regions where their constituents have relatives as a way to curry favor with voters. During relief operations in response to Hurricane Mitch,

---

[4]UNHCR (1997), pp. 46–47.

[5]Woodward (1995), pp. 363–367.

[6]Aid workers were aware of the problem at the time but felt their reason for being required them to provide life sustaining assistance to the camps.

[7]Menkhaus (1998), p. 56.

for example, politicians throughout the United States pressed the military to transport their constituents' donations even when these were of little immediate use in the crisis area.

These demands are the norm, not the exception, during a humanitarian intervention. As a result of these political concerns, the military may be under pressure to unload highly visible shipments of food and medicine even though improving local airfields or supplying forklifts may be a more sensible priority for ensuring the steady flow of relief supplies.

The United States and its allies may also have to work through local governments, even if corrupt, for political as well as practical reasons. In many instances, the most suitable airfields or other logistics essentials may be in states near the crisis region, which are only willing to cooperate with relief efforts for a price. Intervening forces may also want regional states to contribute troops or other assets to the overall mission. In general, the political imperative of maintaining good relations with regional states outweighs the occasional humanitarian benefits of bypassing these states in the relief effort. Thus, military forces may have to cooperate closely with regional states even when this contributes little to the immediate mission.

Reluctance to Risk Casualties

The United States is reluctant to risk casualties during complex contingency operations. This reluctance is understandable considering that these operations seldom involve a vital interest, but it constrains military commanders and overall operations. Preventing casualties is often a higher priority for the military during complex contingency operations than humanitarian objectives. Political leaders are highly sensitive to *any* casualties. Due to the limited political will common to these operations, casualties can lead to an intervention's abrupt termination or the curtailment of many of its activities. Even a small number of casualties may cause the premature withdrawal of military forces.

To take an oft-cited example, the United States had a limited geopolitical interest in Somalia during the Cold War, when it contended with the Soviet Union for influence on the Horn of Africa. After the

Cold War ended, the United States lost its geopolitical interest in Somalia. The Clinton administration tried to pass operational responsibility to the United Nations but still became deeply involved in a struggle with one of the clan leaders. When U.S. forces suffered 18 deaths on October 3, 1993, the administration could not justify these losses or even explain its policy to the satisfaction of Congressional critics. In response, the administration reinforced the deployed forces but reduced their mission and announced a firm date for departure.

Reluctance to risk casualties had a strong influence on the course of recent events in Kosovo. Thinking more of their own constituencies than their adversary, the NATO powers announced publicly that they would not attempt to enter Kosovo without permission from the Yugoslav government. On March 24, 1999, they began Operation Allied Force as an air-only campaign to coerce Belgrade while minimizing their own casualties. In late May, major NATO powers started a public debate on the use of ground forces, with Britain in favor, Germany opposed, and the United States and Italy willing at least to consider the option. In congressional testimony, U.S. officials repeatedly stressed that there was no consensus in NATO for a ground invasion of Kosovo, despite brutal "ethnic cleansing" of Kosovar Albanians by Yugoslav forces.

Concern for casualties may also lead to restrictions on where troops can go, which areas receive aid, and the types of military activities conducted. Because of concerns about casualties in Rwanda, U.S. soldiers were limited in areas they could deploy because of ongoing fighting. Those areas, in turn, received only limited aid because of the dangers of sending food to the region. In addition, political leaders are generally hesitant to assign potentially dangerous tasks to the military, such as disarming or separating combatants, even if these are essential to mitigating a conflict.[8]

As a result of casualty sensitivity, force protection is a priority even when no peacekeeping activities are planned. When a government is

---

[8]In Rwanda, the head of UNHCR, Sadako Ogata, called for member states of the United Nations to use force to separate the "genocidaires" from the legitimate refugees, but no government was willing to take on the dangerous and difficult task. In the end, UNHCR settled for paying Zairian army troops to keep order in the camps.

absent or hostile, intervening powers must anticipate that soldiers could be caught in local conflicts or deliberately targeted by warring parties. In Support Hope, the military assessed that the civil war would probably resume, and that this would make safety a constant concern of U.S. personnel. In the Goma refugee camp, armed paramilitary forces roamed the area, and movement after dark was forbidden. Local Zairian forces did more harm than good, acting as bandits rather than assisting law enforcement.[9] As the U.S. Air Forces in Europe (USAFE) after-action report on Support Hope noted, "The boundary between peace and war in these conditions is vague and easily crossed; corrupt government troops, mobs and bandits in paramilitary uniforms are the likely threats."[10]

The focus on force protection also leads to tension between relief agencies' long-term expectations and political realities. Relief agencies often want the military to provide a secure environment for the long term. Thus they use their political and media influence to press military forces to enforce a cease-fire, disarm combatants, and otherwise undertake potentially dangerous operations. The military, however, may be tasked to avoid such operations because of skepticism about their utility and concerns that they might lead to casualties.

Restrictive Rules of Engagement

During complex contingency operations, the military will usually have to operate with restrictive rules of engagement (ROE). It will typically be allowed all measures necessary for self-defense, but it may be severely restricted with regard to offensive actions, even in response to severe provocations.

These limits on the use of force stem from many sources. Casualty concerns may lead to restrictions on escalating potential conflicts. IOs and NGOs tend to see deadly force as counterproductive. They also fear that the use of force could compromise their impartiality in the eyes of local parties, which may associate U.S.-based relief agencies with U.S. government actions in the region. Thus, their

---

[9]U.S. European Command, *Operation Support Hope*, pp. 4 and 12.

[10]U.S. European Command, *Operation Support Hope*, p. 32.

own personnel may be subject to reprisals. Domestic constituencies may recoil from the use of force, especially during operations to protect humanitarian assistance. Naive perceptions of humanitarian aid may prompt a domestic political backlash if deadly force is used to ensure its delivery.

Belligerents and even civilian populations will quickly grasp the importance of ROE, divine their content, and attempt to exploit them. In Somalia, for example, clan-based militias knew that U.S. forces were severely restricted in applying deadly force when it could endanger innocent civilians. They tried to exploit these ROE by firing from the protection of crowds, using their own civilians as human shields.[11] During the Bosnian conflict, all parties exploited UNPROFOR's overly elaborate and highly restrictive ROE, especially as concerned close air support.[12]

The United States and its allies have learned from past problems, however, and ROE in future operations are likely to be better designed and more flexible than in the past. In Bosnia and Kosovo, NATO learned from the UNPROFOR experience and put the former belligerents on notice that its own ROE were sufficiently robust to enable NATO forces to use necessary force to protect themselves and keep the peace. In future deployments, sufficiently robust ROE are likely due to increased political recognition of their importance.

BALANCING CONFLICT AND RECONCILIATION

Many of the constraints on military operations also stem from the nature of the crises themselves. Military forces may find it difficult to

---

[11]The United States responded by using snipers to engage the Somali gunmen, but in some situations it relaxed the ROE. During the fighting on October 3, 1993, for example, U.S. forces delivered heavy fires in sections of Mogadishu to protect special operations forces, which were surrounded by supporters of Aideed.

[12]"When UNPROFOR arrived in Bosnia the locals mentally paused to assess what impact UN forces would have on their country. They expected that the UN would make a big difference. It had some impact but overall UNPROFOR appeared to be ineffective. So-called experienced peacekeepers applied their methods too rigidly. Therefore, after UNPROFOR's first arrival, the Bosnian conflict carried on much as before, with the United Nations' forces simply being regarded as an annoyance that was sometimes in the way." Stewart (1993), p. 326.

maintain impartiality. More fundamentally, their actions often have only a limited impact on the dynamics of a conflict.

The Limits of Impartiality

The penchant for impartiality often limits military action. For operations authorized by the Security Council, impartiality means that the Council does not recognize a specific aggressor and holds all parties responsible for the conflict. Impartiality is often essential to peace operations, including the protection of humanitarian assistance. Impartiality, however, often imposes constraints on military forces that hinder their effective use.[13]

When impartiality has been declared, intervening forces are prohibited from assisting any one party, even when it appears to be a victim. Thus, UNPROFOR was not allowed to assist Muslims in their defense of Sarajevo, even when Bosnian Serbs were egregiously violating Security Council resolutions with respect to the city. Military forces are also prohibited from sharing information that could have intelligence value with belligerents during an ongoing conflict. Sometimes the effort to maintain impartiality takes strange forms. During Operation Allied Force, for example, NATO spokesmen and U.S. officials denied that NATO was the KLA's air force, although to some extent it did play that role. Subsequent reports revealed that Serb forces were considerably more vulnerable to air attacks while responding to the KLA, even when the KLA operations were not successful in and of themselves.

In practice, impartiality is difficult to attain. Simply by helping enforce a peace and provide humanitarian assistance, intervening

---

[13]Impartiality is frequently confused with neutrality. Neutrality implies that no military force may be used. Impartiality implies that military force may be used only against parties that violate agreements or resolutions of the Security Council, assuming that the operation is coercive. Impartiality further implies that enforcement will be even-handed, i.e., directed equally against all violators. In Somalia, for example, the Special Representative of the Secretary General sought to apprehend Mohammed Farah Aideed because he was in violation of agreements concluded with the other parties. Through this action, the Special Representative did not cease to be impartial, assuming that he was equally willing to apprehend any other violator. Of course, the targets of punitive action usually claim, as Aideed did, that the United Nations and its agents have unfairly singled them out.

powers are aiding some parties to a conflict more than others.[14] Intervening forces can try to minimize their impact on the local balance of power, but warlords and other local parties will be acutely aware of any impact, no matter how benign the intention.

Limits to the Utility of Force

The warring parties' acceptance of terms, often in the form of a peace plan, is the first step, not the last. The ultimate goal of many complex contingency operations is reconciliation of the parties. This ambition, however, cannot always be accomplished through military operations, particularly when military resources are limited and political will for an operation is weak.

Reconciliation can take different forms. In Somalia, it implied that the warring clans would submerge their differences in common support of a democratic central government, which no single clan or group of clans would dominate to the detriment of others. In Haiti, it implied that impoverished masses led by Aristide and former supporters of the Duvalier family would join in creating the country's first successful democracy. In Bosnia, it implied that three ethnic groups, bitter enemies after four years of fighting and "ethnic cleansing," would live together harmoniously in one multiethnic state. In Kosovo, it currently implies that a large majority of Kosovar Albanians and a small minority of Serbs will become contented citizens of an autonomous province within Serbia, although Serbs recently perpetrated widespread atrocities against Albanians. Reconciliation may take years, if not generations, to accomplish and is not achievable through military force.

Although the military can stop a war, by itself it cannot bring a lasting peace. In the deepest sense, peace means more than cessation of hostilities. It means resolution of those problems that led to conflict and could cause conflict again if they remain unresolved. For example, a UN-controlled force has been in Cyprus since 1964. Since the Turkish invasion in 1974, it has monitored a buffer zone between the cease-fire lines. In recent years, the force has dwindled to three small battalions, financed in part by the governments of Greece and

---

[14]Betts (1994); Seybolt (1996).

Cyprus. The force undoubtedly inhibits outbreak of armed conflict by separating the parties, but it also helps them to evade settling their differences and bringing peace to the island. The zone of separation in Bosnia may have similar longevity, illustrating both the utility of military force and its limitations.

The military may suffer a gap between declared political aims and the means available to it. The occasionally ambitious nature of U.S. objectives, the limits on U.S. and allied political will, and the complex and often intractable nature of many of the conflicts often place the United States and its allies in a position of seeking difficult or resource- and time-intensive goals with relatively few means. The military can often help make progress on humanitarian goals, but it can seldom solve many of the underlying problems by itself. Moreover, its mission and resources are likely to be limited in both duration and extent.

The most glaring example is the U.S. intervention in Somalia. Under UNOSOM II, the United States was tasked to promote nation-building and promote reconciliation in Somalia. Such an ambitious agenda, however, would have required a large, long-term deployment, considerable financial aid, and other types of intervention that the United States and its allies did not favor. Indeed, military officials made clear that forces in Somalia were not adequate to the ambitious mission at hand. Although the collapse of the reconciliation effort in Somalia and the deaths of 18 U.S. soldiers have made both political leaders and military officials more aware of such gaps, they are often difficult to avoid completely.

ADVANCED PLANNING DIFFICULTIES

Because of the political nature of interventions and the large number of actors involved, advanced planning with NGOs and allies—particularly finely tuned operational planning—is often difficult, complicating coordination in the early days of a crisis. The military often has little warning time to prepare. Thus, it may be unaware of ongoing relief agency activities in a crisis area or of agencies' plans for intervention.

Operation Support Hope typifies such interventions. In Support Hope, deployment and execution occurred almost simultaneously.

Collecting, evaluating, and disseminating information was done "on the fly." The United States changed the location of one of the JTF headquarters from Kigali in *midair,* because it had decided for political reasons that it would not recognize the new government of Rwanda immediately.[15] The United States and its European allies did not coordinate their efforts initially. France in particular often worked at cross-purposes with Washington.

From an NGO point of view, advanced planning with the military is of only limited value. Because the military's participation is unpredictable and usually short-term, investing scarce NGO time and resources into better relations often provides little benefit during actual crises. Several NGO interlocutors noted that their efforts over the past several years were "wasted" due to a lack of military participation in providing relief.

Not all the above constraints can be overcome—many are inherent to these types of operations. Nevertheless, by improving coordination with European allies and relief agencies—the emphasis of the remainder of this report—many of the constraints can be minimized, greatly improving the effectiveness of the relief effort.

---

[15]U.S. European Command, *Operation Support Hope*, pp. 3 and 12.

PART TWO. THE RELIEF COMMUNITY AND THE MILITARY

The challenges inherent to complex contingency operations cannot be completely overcome, but they can be reduced. One high-payoff area is for the military to improve its cooperation with relief agencies—the focus of the following four chapters. NGOs and IOs play an essential role in relief operations, and the military has been most successful in the past when it has successfully supported these organizations, using its own capabilities to bolster theirs. Such support is only the beginning. A better relief agency–military partnership has tremendous potential when anticipating a conflict, allowing all partners to respond to it more quickly and ensure that the response is more effective.

Part Two first provides an overview of the relief community. It then describes possible advantages to better arrangements and notes current coordination structures. It concludes with an assessment of obstacles to coordination, pointing out their sources.

Chapter Six

OVERVIEW OF THE RELIEF COMMUNITY

The first step toward better coordination is for the military to gain a better understanding of relief agencies. The relief community is not monolithic. The actors vary tremendously in their capabilities, size, and attitudes, with considerable implications for cooperation with the U.S. military and for the success of the overall relief effort. Major actors include the United Nations family, the Red Cross and Red Crescent Movement, and NGOs. Understanding these various players is a precondition to coordinating their activities. This chapter identifies major actors within the relief community and categorizes NGOs in a manner that will help the U.S. military understand their various missions and capabilities.

UNITED NATIONS FAMILY

The United Nations is a family of related organizations that includes six principal organs, numerous programs, and specialized agencies. Despite efforts at reform, coordination across these organizations remains poor.

Principal Organs

The United Nations has six principal organs: General Assembly, Security Council, Economic and Social Council, Trusteeship Council, International Court of Justice, and the Secretariat. Three of these organs are especially important during humanitarian interventions: the Security Council, the Economic and Social Council, and the Secretariat.

According to the UN Charter, the Security Council has primary responsibility for maintenance of international peace and security. It expresses its will in resolutions, which must have concurrence (assent or abstention) from all five permanent members (Britain, China, France, Russia, and the United States). The Security Council usually defines the mandates for peace operations by its resolutions. However, member states often act unilaterally or as part of an alliance without approval from the Security Council, as in Operation Allied Force.

The Economic and Social Council (ECOSOC) has broad responsibility to coordinate the economic and social work of the entire UN family. It usually meets in plenum once annually, alternating between New York and Geneva. It routinely consults with NGOs whose work falls under its competence. Currently about 1,500 NGOs hold consultative status with the Council. It sorts these NGOs in three categories: Category I are routinely consulted; Category II have specialized expertise; and Category III are consulted on an ad hoc basis.

The Secretariat, headed by the Secretary-General, includes two entities that are important for relief operations: the Department of Peacekeeping Operations (DPKO) and the Office for the Coordination of Humanitarian Affairs (OCHA). DPKO provides direction and logistic support to UN-controlled peace operations. OCHA is intended to strengthen coordination among UN agencies that respond to emergencies. The head of OCHA is simultaneously the Under Secretary-General for Humanitarian Affairs and the Emergency Relief Coordinator for all specialized agencies.

Programs and Specialized Agencies

In addition to the principal organs, the UN has a wide variety of programs and specialized agencies,[1] most falling under cognizance of

---

[1]A list of some of the major programs and agencies would include United Nations Children's Fund (UNICEF); United Nations Conference on Trade and Development (UNCTAD); United Nations Development Programme (UNDP); United Nations Volunteers (UNV); United Nations Office for Project Services (UNOPS); United Nations Environment Programme (UNEP); United Nations Population Fund (UNFPA); United Nations Relief and Works Agency for Palestine Refugees in the near East (UNRWA); United Nations University (UNU); World Food Programme (WFP); United Nations High Commissioner for Human Rights (UNHCHR); United Nations Centre for

the General Assembly and the Economic and Social Council. They are not subordinated to the Secretariat and therefore need not accept direction from the Secretary-General. They include several organizations that play important roles in relief operations, described below.

The United Nations High Commissioner for Refugees (UNHCR) works on behalf of refugees to secure their protection, provide assistance to them, and seek durable solutions to their problems. It can serve as a lead UN agency, especially in the initial phase of a humanitarian crisis. These solutions might include repatriation, asylum in the country where refugees have fled, or resettlement in a third country. UNHCR maintains an office in New York, but its headquarters is in Geneva. It is advised by a large Executive Committee that meets annually and accepts direction from the General Assembly and ECOSOC. UNHCR has an annual budget of approximately $1.2 billion, derived almost exclusively from voluntary contributions. It currently has over 5,000 employees working in 122 countries, but it works primarily through about 450 NGOs.

The World Food Programme (WFP) is the world's largest multilateral provider of food aid. In contrast to the UNHCR, WFP is focused on logistics and food delivery and does not serve as a lead agency. WFP headquarters is in Rome and its current director is an American with experience in the U.S. Department of Agriculture. A committee, half appointed by ECOSOC and half by the Food and Agricultural Organization, governs WFP. Most aid is donated in kind by member states with U.S. agricultural surplus playing a large role. During 1997, WFP delivered 2.7 million tons of food in 84 countries. To deliver this aid, WFP charters commercial carriers on a large scale.

Human Settlements (Habitat); United Nations High Commissioner for Refugees (UNHCR); Office for Drug Control and Crime Prevention (ODCCP); United Nations Development Fund for Women (UNIFEM); International Labour Organization (ILO); Food and Agriculture Organization of the United Nations (FAO); United Nations Educational, Scientific, and Cultural Organization (UNESCO); International Civil Aviation Organization (ICAO); World Health Organization (WHO); World Bank Group; International Monetary Fund (IMF); Universal Postal Union (UPU); International Telecommunication Union (ITU); World Meteorological Organization (WMO); International Maritime Organization (IMO); World Intellectual Property Organization (WIPO); International Fund for Agricultural Development (IFAD); United Nations Industrial Development Organization (UNIDO); International Atomic Energy Agency (IAEA); and World Trade Organization (WTO).

The World Health Organization (WHO) is headquartered in Geneva and gives guidance in health matters and works to strengthen government health programs. An assembly that includes all member states of the United Nations governs it. Its annual budget is about $800 million.

The Food and Agricultural Organization (FAO) of the United Nations promotes agricultural development and helps countries provide for emergency relief. It has headquarters in Rome and is governed by a biennial conference of member states. It administers approximately $2 billion annually, received as voluntary contributions from government and private donors.

The United Nations Children's Fund (UNICEF) promotes children's rights and supports programs that increase their well-being. It reports to the General Assembly through ECOSOC. It has an annual budget of approximately $0.9 billion, derived from voluntary contributions. It employs about 6,200 persons in 133 countries and has several main offices.

The United Nations Development Programme (UNDP) funds programs for sustainable development and coordinates technical assistance. UNDP has its headquarters in New York and 132 offices worldwide. It is governed by an Executive Board, which represents developed and developing countries. It concentrates its efforts in the world's poorest countries. It has an annual budget of approximately $700 million from voluntary contributions and administers another $1.4 billion annually from a variety of special funds.

RED CROSS AND RED CRESCENT MOVEMENT

The Red Cross and Red Crescent Movement straddles the gap between international organizations and NGOs. It is a private organization independent of all international organizations and governments, yet it has official status through treaty, agreement, and usage. The movement includes the International Committee of the Red Cross (ICRC), the International Federation of Red Cross and Red Crescent Societies, and affiliated national societies.

International Committee of the Red Cross

The ICRC, the most important partner for the military in overseas humanitarian crises, is quite distinct from NGOs and UN agencies and is effectively in a class by itself as a relief provider. The ICRC has an international mandate to promote compliance with humanitarian law and to help victims of conflict. It receives funding from many governments, especially the U.S. government, and from nationally based Red Cross and Red Crescent societies. It administers an annual budget of about $550 million. It has about 650 personnel in its Geneva headquarters and about 7,800 personnel worldwide, the overwhelming majority of them locally hired. It maintains a permanent presence in more than 50 countries.

The ICRC is formally tasked by the Geneva Conventions of 1929 and 1949, which concern humane treatment of prisoners of war and civilian victims of conflict. To carry out its tasks, the ICRC must have freedom of movement within areas of armed conflict and across lines of confrontation. It ensures this freedom by being completely independent, impartial, and neutral. The ICRC reminds authorities involved in armed conflict of their obligations under international law to observe certain rules of conduct. In addition to its monitoring functions, the ICRC distributes relief supplies, provides emergency treatment, and administers care for the disabled.

To accomplish its mandate, the ICRC must be able to cross lines of confrontation and move freely throughout areas of conflict. It cannot expect to have this access unless belligerents are persuaded that the ICRC is neutral and impartial. *Any* indication that the ICRC may have intentionally or even inadvertently aided one side in a preferential fashion may destroy this reputation and serve as a pretext to limit its access. Understandably, the ICRC is anxious to preserve an unblemished record for neutrality and impartiality.

International Federation

The International Federation of Red Cross and Red Crescent Societies promotes affiliated national societies and gives unity to the movement. It is governed by a General Assembly of all National Societies that meets biannually.

National Societies

National Societies exist in most of the countries of the world. They focus on the well-being of their specific nations. The American Red Cross has an International Services Department with an annual budget of $20–$30 million that channels relief through the ICRC and the International Federation.

NONGOVERNMENT ORGANIZATIONS

NGOs are voluntary associations independent of government control that seek to realize human rights and to provide humanitarian assistance according to need. By one conservative estimate, there are more than 26,000 NGOs that operate in more than one country, and several million more exist inside national borders.[2] Just about anybody anywhere in the world can establish an NGO if he pleases. As a result, there are thousands of NGOs and their composition fluctuates constantly. For example, 18,000 NGOs attended a parallel forum to the Rio Conference on the environment, and 1,400 were formally registered with the conference itself. NGOs vary widely in terms of their capabilities, professionalism, and willingness to cooperate with military forces, including the U.S. military. Successful coordination requires understanding these differences.

UN agencies and national governments often rely on NGOs as integral parts of their relief and development efforts. The European Union channels more than half of its aid through NGOs. Similarly, the WFP and the UNHCR rely heavily on NGOs to run refugee camps, deliver food, and otherwise conduct essential missions. By some measures, NGOs have surpassed the World Bank in dispersing money.[3]

The skills and size of NGO personnel vary tremendously.[4] There are NGOs with a staff of thousands that supplement a core of competi-

---

[2]"The Non-Governmental Order" (1999), p. 21.

[3]"Sins of the Secular Missionaries" (2000), p. 25.

[4]Undifferentiated generalizations about NGOs are not very useful because the community is so diverse. CARE and a local civic organization are both "NGOs," just as the U.S. armed forces and, say, the Army of Luxembourg (one light infantry battalion) are both "militaries."

tively salaried professionals with unpaid volunteers, and NGOs that literally consist of one individual and a few friends. The four largest receivers of U.S. government funding—Cooperative for Assistance and Relief Everywhere (CARE), Catholic Relief Services (CRS), Save the Children, and World Vision—are skilled, dedicated, and able to participate in a wide range of humanitarian relief activities. Médecins Sans Frontières (MSF) (Doctors Without Borders), Oxfam, and several other international NGOs are highly capable and competent. At the opposite extreme are small organizations composed largely of relief amateurs, which may spring up overnight. In Rwanda, for example, a woman named Ruth formed an organization aptly entitled "Ruth Cares"—a single individual with no appreciable skills other than a desire to help. Often these groups involve concerned citizens in the United States. As another example, the Defense Department once helped some upstate New York women send sewing materials to South Africa to support a sewing club.

NGOs also vary by issue area, specialization, geographic coverage, and degree of institutionalization:

- NGOs focus on a wide range of issues, including natural disasters, refugees, underdevelopment, the environment, and child labor. Some NGOs define themselves primarily by an issue (the environment, women's affairs, children's rights, health, agriculture, animal rights, political prisoners, famine relief, recovery of avalanche victims, and so on), others by ideology (Third World solidarity, pacifism, etc.), by sympathy for a specific country, ethnic group, or region (e.g., immigrants from Central American countries helping victims of natural disasters in those countries), or by religious charity (Christian, Jewish, Islamic, etc.).

- Some NGOs consist of members of one profession only (e.g., health professionals or members of one medical specialty such as dentists or ophthalmologists) who may either work for that organization full time or donate a few weeks of pro bono work each year to go on a mission. Other NGO staff members have no skills as such but concentrate on collecting used clothing, food items, and other donations.

- NGOs may represent their club, their city, their locality, their ethnic group, their country, their continent, or their religion; or they may simply be international. They can be affiliated with their local, state, or national government, their church, or some other organization, or they can eschew affiliations.

- NGOs can be designed as permanent organizations or be dedicated to one conflict only, such as the organizations that sprang up in response to the Bosnian conflict.

The above variables do not necessarily correlate with the success and prestige of an organization. Staid organizations such as the European Catholic monolith Caritas number among the big players, but organizations that use unconventional and sometimes drastic means of protest, such as Greenpeace, or that take a decidedly radical political stance, such as Médecins Sans Frontières also enjoy widespread public support and respect.

As a general rule, we can expect the financially more powerful, more reputable organizations with good media ties and an experienced staff to be more important in any given locale, but there are important exceptions. In emergency situations, size, experience, and reputation are not the only predictors of value. An obscure missionary organization may find itself in possession of the only functioning aircraft for a critical 48-hour period; a hitherto unknown group running a remote clinic might have the only available cartons of vaccine; a small partisan organization with a friendly relationship to a local warlord might be the only quick source of information about events in a particular region or safety guarantees to access it.

NGOs also vary widely in the types of aid they provide. Some are concerned with immediate assistance, some with long-term development, and some address both areas. In recent years, several of the larger NGOs appear to be devoting greater resources to immediate assistance.[5] Some NGOs specialize by region or by functional area. The Catholic Medical Mission Board, for example, provides emergency and long-term health care worldwide, while the American Refugee Committee helps care for and train refugees. Medical Care Development, Volunteers in Technical Assistance, and Africare all focus their relief efforts on Africa, while other NGOs are active in several regions or worldwide. In the developing world, some NGOs

---

[5]Within major NGOs, individuals concentrating on short-term relief are often gaining more influence and resources than those concentrating on long-term development.

are government-run and may be highly politicized. In Kosovo, for example, the Albanian government ran an NGO that furthered its own policies in the months preceding the 1999 NATO air campaign.

Although a small number of well-established NGOs contribute most of the overall effort, hundreds of NGOs may operate in a region and cannot be safely ignored. Personnel from small NGOs may require disproportionate attention from the military if they attempt to cross lines of confrontation or operate in areas of intense conflict. Moreover, a small NGO may have an influential domestic constituency to whom it can plead its case, enabling it to exert political influence on the overall relief operation.

NGOs also differ widely in their ability and willingness to cooperate with the military. In general, European NGOs tend to avoid close cooperation, while U.S. NGOs are more amenable. But even well-established U.S.-based NGOs differ in the degree to which they will openly associate with the U.S. military. Some welcome closer ties; others fear that their impartiality might be compromised. Some religiously affiliated NGOs eschew the use of force and regard the military with suspicion. For example, Quaker and Mennonite NGOs have a long tradition of nonviolence and are highly reluctant to endorse any use of force.

Several large NGOs are trying to improve cooperation with the military. For example, CARE and World Vision have hired former military officers to facilitate better cooperation. In ideological terms, the end of the Cold War has made such cooperation more palatable. While the Cold War was still in progress, many NGOs hesitated to cooperate with the U.S. military because it stood on one side of an ideological divide. More importantly, NGOs have grown increasingly concerned about security for their personnel in such places as Somalia, Bosnia, and Kosovo, making them more welcoming of security that the military can provide.

CATEGORIZING NONGOVERNMENT ORGANIZATIONS

It would be practically impossible to coordinate with all NGOs during an emergency if each demanded the same amount of attention. Coordination becomes a more manageable problem if NGOs are categorized in a useful way, enabling the military to determine which

NGOs are most likely to cooperate and which have the most to offer in any particular operation. We suggest the following taxonomy, which will enable the military to concentrate its resources accordingly:

- Core-Team: highly competent, broadly capable, and predisposed to cooperate with the military.

- Core-Individual: highly competent and broadly capable, but less eager to cooperate with the military.

- Specialized: highly competent and capable in select functional areas.

- Advocacy: dedicated to promoting human rights but not normally providers of material assistance.

- Minor: competent but having less capability than the core-team.

Core-Team

The core-team NGOs devote appreciable portions of their resources to immediate relief of suffering. Most of them receive substantial support from the U.S. government, including grants, contracts, and in-kind transfers. Most of them work closely with the Office of Foreign Disaster Assistance (OFDA) of USAID to coordinate the U.S. response to an emergency, both in Washington and in the field. During emergencies, OFDA may invite these organizations to send representatives to its operations center. Several of these NGOs send representatives to participate in conferences, seminars, and exercises sponsored by the military. During interviews, officers from these organizations expressed willingness to cooperate more closely with the military.

Taking as a threshold a gross annual revenue of $30 million or more,[6] the following organizations fall into this group: Adventist Development and Relief Agency (ADRA), Africare, American Jewish

---

[6]Total annual revenue as reported in Geoghegan and Allen (1997). In the highly competitive world of NGOs, high revenue generally indicates strong capability. It implies that an NGO has maintained a level of performance over time that attracts donors, including in most cases the U.S. government.

World Service (AJWS), American Red Cross (International Services Department), Cooperative for Assistance and Relief Everywhere (CARE), Catholic Relief Services (CRS), Church World Service (CWS), International Aid, International Rescue Committee (IRC), Mercy Corps International (MCI), Save the Children (U.S. chapter), United Methodist Committee on Relief (UMCOR), and World Vision Relief and Development (WVRD).

Core-Individual

These organizations are international and most of their assets are located outside the United States. Although they often receive funding from the U.S. government, they display highly independent attitudes. They may reject support from the military or criticize the military in strong terms, even while accepting its support. Their criticism might include allegations that the military is obsessed with self-protection, insensitive to cultural differences, and disruptive to already established patterns of aid. It might also include allegations that the military is taking sides unnecessarily or being used to pursue political goals that will not allay or might exacerbate the conflict.

Two international relief organizations falling into this group are Médecins Sans Frontières and Oxford Committee for Famine Relief (Oxfam). Both organizations receive U.S. government aid but strive to maintain their distance from the U.S. government's agenda. Médecins Sans Frontières was founded in 1971 by French doctors who wanted to provide medical assistance during emergencies completely independent of political, religious, or economic considerations. It provides not only medical care and training but also limited humanitarian assistance of other kinds. Its medical personnel are highly skilled in emergency medical care, immunization, sanitation, and basic hygiene. In addition, NGOs may speak out against violations of human rights they observe during their work. Oxfam was founded in England during 1942 to address suffering caused by the war. It provides emergency relief and also carries out programs to promote long-term development.

Core-individual organizations are not opposed to all coordination with the military, but their ideals and preferences often make planning and sustained coordination more difficult. These NGOs at times will accept and even request U.S. military assistance. As institutions, however, they will try to avoid open identification with the United States, particularly the U.S. military. As one MSF official noted:

> We try as much as we can to differentiate from any military that is present. The image of cooperating with the Air Force is scary for us. This would mean recognizing that we are part of the conflict, and it would send a confusing message to the populations we are trying to help.

As with all NGOs, the particular response of a core–individual NGO will be shaped by local circumstances and the individuals involved.

Government support, even something as limited as accepting stipends to pay for conference attendance, usually provokes much agonizing and soul-searching among NGO officials. Offers of significant funding can often be turned down on the chance that it might make the organization appear partisan or dependent. MSF policy requires at least half of all funds to come from private sources, and "has shied away from French government funding."[7] Yet these requirements are often honored more in the breach. Oxfam received roughly a quarter of its 1998 budget from the British government and the EU; MSF received 46 percent from various governments.[8]

Specialized

Some NGOs lack the broad capabilities of the core organizations but are highly competent and capable in functional areas, such as emergency medicine. Awareness of their capabilities is vital, as they can be useful in certain kinds of crises. Often, they are more important than core NGOs when a crisis falls into their area of specialty.

Such organizations include Agricultural Cooperative Development International/Voluntary Overseas Cooperative Assistance (ACDI/VOCA), Action Against Hunger, African-American Institute (AAI), American Refugee Committee (ARC), The Brother's Brother Foundation (BBF), Catholic Medical Mission Board (CMMB),

---

[7]Brauman and Tanguy (1998).

[8]"Sins of the Secular Missionaries" (2000), p. 25.

Childreach, Christian Children's Fund (CCF), Direct Relief International, Food for the Hungry International (FHI), Heifer Project International (HPI), MAP International, Medical Care Development (MCD), Winrock International, World Relief Corporation, and the U.S. Young Men's Christian Association (YMCA).

ACDI/VOCA provides technical expertise to business and government agencies. Action Against Hunger is the U.S. arm of an international organization known in France as *Action Contre la Faim* that specializes in disaster relief. With strong support from the U.S. government, AAI conducts exchange, information, and conference programs in Africa. ARC works to ensure survival of refugees and displaced persons. BBF promotes international health and education by distributing donated resources. CMMB provides emergency health care and conducts longer-term programs to make health care available to impoverished people. Childreach strives to assist needy children through sponsorship. CCF works to protect children and promote their development. Direct Relief International provides emergency medical supplies and shelter to victims of disaster and also conducts training for medical personnel. FHI provides food and material aid for disaster victims. HPI specializes in providing income-producing livestock. MAP International provides emergency medical care and distributes medical supplies. MCD designs and implements programs to provide emergency relief and promote public health. Winrock International works to increase agricultural productivity. World Relief Corporation provides disaster relief on behalf of evangelical churches. YMCA focuses on education and vocational training.

Advocacy

Advocacy organizations promote human rights or other goals but do not normally provide material assistance. Examples include Amnesty International, Immigration and Refugee Services of America (IRSA), International Center for Research on Women (ICRW), Physicians for Human Rights (PHR), Refugees International (RI), United States Catholic Conference (USCC), and United States Committee for Refugees (USCR).

Amnesty International advocates observance of human rights as set forth in the Universal Declaration of Human Rights. IRSA promotes fair and humane public policy concerning people in migration. ICRW raises awareness of women's contribution to development. PHR uses forensic science to investigate violations of human rights. RI seeks to bring the plight of refugees to the world's attention. USCC advocates policies to address the needs of migrants and refugees. USCR defends the rights of refugees, asylum seekers, and internally displaced persons.

Although advocacy organizations often are of little immediate utility during relief operations, the military cannot afford to ignore their needs or activities. These agencies may play a key role in shaping U.S. political objectives and domestic opinion on the efficacy of the relief effort. Moreover, they often have strong grassroots components to gain political support for their objectives.

Minor

Minor organizations may or may not be competent in providing relief. They range from organizations with substantial annual revenues ($5–$30 million), which can make strong contributions in certain fields, to much smaller organizations, which can make only small contributions. Most NGOs fall into this category. Although minor organizations contribute little materially when compared with core and specialized NGOs, they can play important roles in a particular country or during a particular crisis. Some have political connections or may create problems on the ground because of their activities. Because of their small size, NGOs in this category may not be well known, even to specialists, prior to a particular crisis.

The above typology is not exact and members in each category vary considerably by country and region. Nevertheless, understanding the different capacities and inclinations of NGOs is useful in helping the military employ its scarce resources. The discussion in subsequent chapters draws on this typology when discussing problems and noting possible solutions.

Chapter Seven

ADVANTAGES TO BETTER COORDINATION
WITH THE RELIEF COMMUNITY

Working well with NGOs is essential for the effective provision of relief.[1] These organizations have expertise in rapidly responding to crises, identifying needs, distributing aid, providing essential services, and promoting long-term development. They provide a range of capabilities and skills, many of which are not found or are rare in the military. Moreover, NGOs are usually present in the country before a crisis begins and will usually remain after it ends. By coordinating more effectively with NGOs and other relief agencies, the military can capitalize on their expertise and capabilities to respond more effectively during humanitarian crises.

MORE RAPID RESPONSE

NGOs can react rapidly during a crisis. Often, NGOs respond well before national governments do, and they can quickly move people and small amounts of supplies to trouble spots. As one (Marine) interlocutor noted, "NGOs are more expeditionary than the Marine Corps." After tropical cyclone Marian, NGOs cooperating with Operation Sea Angel quickly identified needs and procured supplies locally, enabling the military to more effectively assist locals.[2] Similarly, in Somalia, Rwanda, and elsewhere, several NGOs were often on the scene in the early days of the crises. From a military point

---

[1]Joint Chiefs of Staff (1996), Joint Pub 3-08, p. III-25.

[2]Seiple (1996), p. 66.

of view, relief agency assets are often the first on the ground, making them useful for assessment, aid distribution, and other essential tasks.[3]

Many NGOs can draw on existing development or missionary networks in a crisis. ADRA International, for example, draws on local Adventist churches in Africa in a crisis; they have prearranged Disaster Assistance Response Team (DART)-like teams from neighboring countries ready to assist on short notice. Other NGOs accustomed to providing education or training farmers in the region also can quickly identify transportation assets, sources of relief supplies, and other essentials during a crisis.

SMOOTHER AIRLIFT

From a USAF point of view, improved coordination can help ensure the proper shipment and distribution of aid during a crisis. After the genocide and refugee crisis in Rwanda, the initial days of the airlift were a debacle. Seventeen countries, 16 NGOs, and several UN agencies simply chartered their own planes and arrived at airfields in the region, assuming that there were personnel and facilities equipped to handle their cargo. Planes landed at airports with limited ramp space and insufficient fuel. Nonpriority cargo often landed before essential items, such as water treatment equipment, because of poor planning. The problem was not limited to non-U.S. assets. Many short-notice requests came from U.S. strategic airlift authorities.

In Africa, the problem of airlift coordination can be particularly acute. African airfields are often small, with a maximum on ground of one or two. Maintenance capabilities may be limited, and fuel scarce. Communications equipment may be lacking or obsolete and ground crews unavailable for unloading. Material-handling equipment for unloading is often absent, limited, or in dismal condition. Moreover, U.S. knowledge of many airfields in the region is incomplete.

---

[3]NGOs, of course, do not respond rapidly to all crises. When a crisis is sudden and massive, NGO capabilities are often strained. Moreover, NGOs are not prepared for particular types of crises, such as environmental disasters (e.g., the Bhopal plant in India).

Politically, ensuring a smooth airlift is often complicated by host-nation sensitivities or poor capabilities. Many African governments are not willing to set priorities for relief (or, at times, care little whether one segment of the population is hungry or in need). Few have advanced air traffic control capabilities. Governments may try to block information about crises for political or bureaucratic reasons. Moreover, African governments are sensitive about sovereignty and reluctant to allow the United States, the UN, or another country to run their airfields.

The USAF could make an important contribution in these crises by coordinating airlift. Although NGOs and UN agencies are often expert at the routine shipment of goods, they lack the ability to conduct and coordinate a major airlift on short notice.[4] This problem becomes immense when, in major operations such as Rwanda, there are over 60 flight carriers and over 100 NGO, IO, and military organizations involved in the relief effort. When the USAF can organize an airlift, however, situations vastly improve. The USAF, through planning and on-time arrivals, can maximize the use of ramp space and prevent planes from landing at airfields that lack fuel. Toward the end of the Rwanda operation, an USAF-run air coordination cell was established in Geneva, improving the airflow.

The problem of airlift coordination is particularly acute in the initial phase of a crisis. Over time, wrangling among donor countries, within the United Nations, and among NGOs is resolved, and they come together in an ad hoc manner. Experience accumulates and procedures are ironed out. This time frame, however, may be measured in weeks or months, while suffering and death continue in the interim.[5]

MORE TIMELY IDENTIFICATION OF LOCAL NEEDS

Relief agencies are highly responsive to the needs of victims, and their input can make the assessment process far more effective for

---

[4]UNHCR, "Report of the Joint Evaluation of Emergency Assistance to Rwanda" (May 1999); interviews with U.S. military and civilian officials corroborate this point.

[5]United Nations High Commissioner for Refugees, *Review of UNHCR Logistics Policies and Practices* (1992).

donor countries and the U.S. military. Relief agency personnel are often well acquainted with the particulars of any crisis. Moreover, some relief agencies also are useful sources of information for the military. Because they are often in-country for many years, they understand the sensitivities of the local culture and the immediate needs of the populace. For example, in much of Africa yellow corn is fed to animals; distributing U.S. yellow corn, although it is widely consumed in America, can cause an affront if supplied as assistance in Africa. Relief organizations are aware of such important cultural preferences. In addition, relief agency officials often live next to the peoples in question and employ large numbers of local nationals, giving them excellent local sources. NGOs at times have access to individuals who, for whatever reasons, will not deal with the military, or that the military does not wish to deal with, or they are in a place where the military has no presence. NGOs are keyed into gossip and information networks. They have language skills and personal links in the community through friendship, personal origin, ethnicity, or marriage. They may also know the local security situation and understand local political realities better than other observers. At times, they have a good sense of what is about to happen or whether a certain approach can work. As Colonel Schultz of the Canadian military noted:

> In September 1996 Ralph Gunhart of UNHCR came to Kingston for the monthly NGO meeting. He had been in Zaire, and he sat there and told us pretty much exactly what was going to happen, three months before it did. I went directly to Intel and asked for a country study; they said we aren't going to do a study because nothing is going to happen there. You've got to listen to the NGOs because they have their finger on the pulse.

In general, many relief agencies are willing to share information with the military on the needs of the local population. Most core-team NGOs recognize the role the military plays and expect to be asked to provide information. Many NGOs are willing to provide information on the local conditions, including the security situation, and to share information before a crisis. This information is valuable in judging the efficacy of the aid effort. Relief agency officials are well positioned to see if the aid is reaching those most in need and to judge the amount being diverted. (Limits to information sharing are discussed below.)

NGO and other relief agencies' knowledge and capabilities, however, should not be overstated. NGO logistics are highly flexible, but they cannot match the overwhelming ability of the U.S. military. Several NGO interlocutors and outside experts criticized NGOs' assertions of expertise, claiming that they often did not understand the big picture. Although a few individual NGO members may be well informed during a crisis, it is difficult to identify them quickly. In addition, many NGOs operate in "surge" mode, providing capabilities and personnel in response to crises. They may know little more of a country's problems than do newly arriving military forces. In Somalia, for example, many NGO personnel arrived after the intervention and knew little about what occurred in-country beforehand.[6] Moreover, NGO personnel regularly rotate (particularly NGO relief personnel, as opposed to those involved in long-term development or missionary work), making for uneven knowledge of an area within an organization. Finally, NGOs have an incentive to exaggerate the extent of a crisis. Because their funding increases when crises become severe, it is in their interest to publicize the most horrifying aspects of a disaster.

BETTER EXPLOITATION OF EXPERTISE

Many NGOs offer expertise in a wide range of relief needs. Some NGOs specialize in sanitation, fighting disease, or providing food, and they employ personnel who have engaged in these tasks for decades or more in a variety of countries. Leading NGOs (and UN agencies such as the WFP and UNICEF) have valuable rules of thumb regarding the food and water needs of civilian populations— knowledge the military has for combat operations but not for humanitarian missions. Equally important, NGOs are often experienced at providing food and medical care to large numbers of people in developing world environments. In the Goma refugee camp in Zaire (Congo), for example, the military initially used IVs to restore body fluids because people were drinking contaminated water. NGOs, however, purified water chemically and simply had people drink, reaching more people with greater speed.

---

[6]Dworken (1995), p. 19.

MORE EFFICIENT USE OF RESOURCES

Relief agencies can reduce the resources the military must devote to an intervention. Because relief agencies are often on the scene before an intervention and have in-place distribution networks and limited stores of food, medicine, and water, the Commander-in-Chief (CINC) can devote his resources to tasks that only the military can provide. Being on the scene after an intervention makes relief agencies ideal partners for "handing off" the humanitarian aspects of the military's mission. In Rwanda, for example, the United States took the lead on providing potable water to refugee camps, but over time the Office of Foreign Disaster Assistance (OFDA) and the UNHCR took responsibility for overall coordination, while the WFP and UNICEF assumed control of food, water, and sanitation.

Relief agencies also are often better able to judge what is a "normal" level of mortality or other measures of effectiveness. If the military is not careful, it may allocate aid poorly and thus perpetuate refugee displacement. In refugee camps in Zaire and Rwanda, the severe poverty of the area made life in the camps seem attractive once the humanitarian aid mission became established. The camps had free food and superb medical care by local standards. As Philip Gourevitch noted, "Zairians who lived in Goma spoke enviously of the refugee entitlements, and several told me they had pretended to be refugees in order to gain admission to camp clinics."[7] NGO knowledge of precrisis conditions can help the military guard against such excesses and resource misallocation.

Relief agencies can also help the military avoid initiating or supporting efforts that cannot be sustained. When the military builds Western-standard hospitals or infrastructure, local governments and authorities may not be able to sustain them once the military leaves.[8] In Somalia, for example, the U.S. military drilled wells to ensure a supply of clean water after the military departed. Maintaining the water pumps, however, proved beyond local capabilities.[9] In past crises in Central America, well-intentioned donors shipped too many

---

[7]Gourevitch (1998), p. 270.

[8]Seiple (1996), p. 11.

[9]Newett (1996), p. 22.

clothes, shutting down local textile industries. NGO knowledge of local cultures and capabilities can help the military build at the appropriate technology level.

NGOs, however, will also sometimes exploit military resources for their own ends during a crisis, leading to inefficient allocations. As one NGO official noted, "NGOs are trained to ask, the military is trained to say yes. So we escalate our demands." For NGOs, demand reacts to supply: Although they might normally purchase goods locally rather than ship them in, if airlift is free they will use airlift regardless of the overall efficiency. Although NGOs can provide insight into overall requirements, they will also exploit military capabilities to get their own supplies in-country.

Chapter Eight

COORDINATION STRUCTURES AND THEIR LIMITS

The large number of disparate actors who may react independently or autonomously make better military coordination with the relief community difficult. Compounding this difficulty is a lack of predictable, dependable control arrangements at the operational level across the United Nations family of organizations and among NGOs. Outside the relief community, the most influential actors create coordination structures, which vary from one operation to another. These may be broadly characterized as host-nation lead, United Nations lead, alliance or coalition lead, and lead country. In addition, the Department of Defense currently funds Centers of Excellence that seek to promote better coordination through a range of initiatives.[1] These structures, however, are often of only limited utility in bridging the gap between international and donor-state objectives and the relief effort on the ground.

The number of disparate actors involved in providing humanitarian assistance complicate efforts to improve coordination. Actors include the relief community outlined in Chapter Six, donor countries, host countries, and regional organizations, displayed graphically in Figure 8.1. At times, everyone and no one may seem to be in charge. Military control arrangements can be highly complex and home governments may micromanage their deployed forces. As a result, the military may not receive entirely clear missions and be

---

[1]Comments on the Center of Excellence in this chapter draw on the experience of the most established center, which is affiliated with USPACOM. A center is also being established by USSOUTHCOM as of this writing.

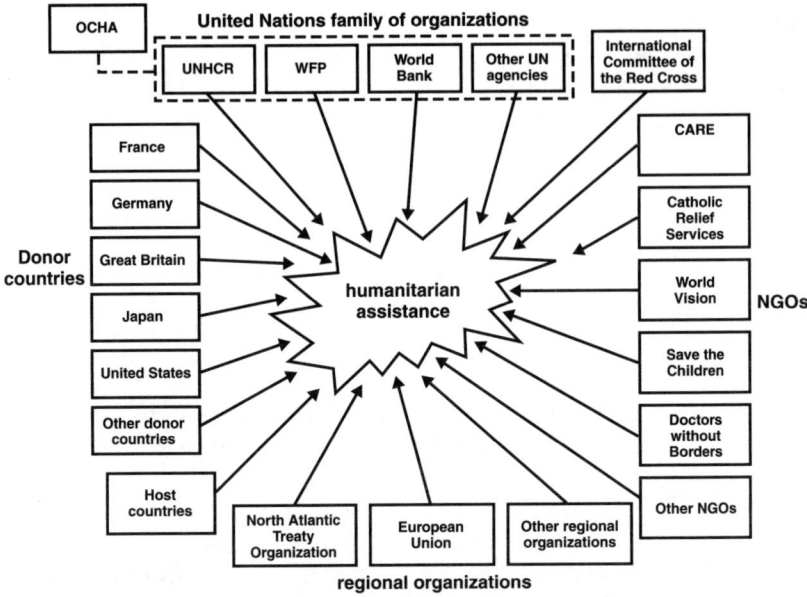

NOTE: OCHA = Office for the Coordination of Humanitarian Affairs; UNHCR = United Nations High Commissioner for Refugees; WFP = World Food Programme.

Figure 8.1—Many Disparate Actors

compelled to improvise, or its mission may change in disconcerting ways.

The major donor countries usually include the United States, European countries (individually and through the European Union), and Japan. These countries may attend donors' conferences, often sponsored or promoted by the United States, where they pledge support to particular efforts. They may contribute without qualification or they may require that their contributions go toward particular geographic or functional areas. The donors may belong to a regional organization, such as the North Atlantic Treaty Organization, the European Union, the Organization for Security and Cooperation in Europe (OSCE), or the Organization of African Unity (OAU), which is directly involved in operations. They may contribute to funding mechanisms such as the World Bank or they may fund individual projects through their national equivalents of USAID. Important donors have bilateral arrangements with host countries, which affect their support and conflict with broader cooperation.

The relief community includes disparate actors that range from the influential UNHCR to small NGOs, some created just to address the particular crisis. Each of these actors makes decisions independently or autonomously. Particularly during the initial phase of a humanitarian crisis, each may pursue its own course of action, subject only to conditions that donors and host countries may impose.

INTERAGENCY PROCESS

Within the U.S. government, complex contingencies may be hampered by a tardy or ineffective interagency process.[2] The departments and agencies of government—especially State, Defense, the U.S. Agency for International Development, Justice, and the Central Intelligence Agency—must all work together, often in unaccustomed ways.

Planning would clearly help, but only the military is likely to hold up its end. In fact, PDD-56 prescribes development of a political-military plan for complex contingency operations, but so far this process has been fitful.[3] The military is familiar with planning and regards the planning process as indispensable, if only because it produces a framework for later improvisation. Civilian departments have often confused plans with schedules and think plans are not worth the effort. Moreover, some officers in the State Department have an aversion to plans, which they see as impediments to the ambiguity and flexibility required for successful negotiation. For example, at the outset of Operation Joint Endeavor in Bosnia, the U.S. military produced a plan to enforce Annex 1A of the Dayton Agreement and was alarmed to discover that no other department had produced a comparable plan.

---

[2]For an analysis of the interagency process in complex contingencies, see Pirnie (1998).

[3]Indeed, one report declares that neither the spirit nor intent of PDD-56 is being followed. Operations in Afghanistan, Bosnia, Serbia, and elsewhere ignored PDD-56 procedures. Scarborough (1999).

Another impediment to coordination is the lack of parallel Department of State and Department of Defense structures on the ground. The Defense Department has regional commands (the unified commands) and regional commanders. The State Department, on the other hand, has ambassadors for each nation but no on-the-ground regional entity whose domain corresponds to that of a CINC. This lack of a State Department regional entity can create confusion by generating multiple reports from the same region and, simultaneously, hinders the development of a coherent presentation of information and responsibilities from the State Department's point of view.

INTERNATIONAL COORDINATION

Fitful as it may be, the U.S. interagency process is a model of efficiency and clarity compared with the international aspects of coordination during complex contingency operations. The arrangements for Bosnia are so complex as to appear unworkable. Indeed, they would be unworkable if the major powers did not share a common understanding of the goals and promote these goals in various venues, including the Security Council, the North Atlantic Council, the Peace Implementation Council, the OSCE, and the Contact Group. The arrangements in Kosovo are similarly complex, although according more formal authority to the Special Representative of the Secretary-General than was accorded initially to the High Representative in Bosnia. In addition, donor countries, the World Bank, and other international financial institutions usually play important roles. Finally, there are a bewildering variety of NGOs, largely funded by the same donor countries but independent of any direct control.

OPERATIONAL-LEVEL ARRANGEMENTS

The relief community suffers from lack of predictable, dependable arrangements to coordinate the United Nations family of organizations and NGOs at the operational level. The concept of strategic, operational, and tactical levels, familiar to military officers,[4] is shown

---

[4]Definitions are contained in joint documents, including *Unified Action Armed Forces*, Joint Publication 0-2; *Department of Defense Dictionary of Military and Associated*

in Table 8.1 for control arrangements for the U.S. government, the U.S. military, the United Nations family, the International Red Cross and Crescent Movement, and NGOs.[5]

Broadly speaking, the UN family of organizations has a formal arrangement for operational-level coordination but fails to implement it in practice. Alone in the relief community, the ICRC is fully operational and controls operations through Delegates General. NGOs have no formal arrangement to ensure operational-level coordination and must find a venue during actual crises.

Coordination Across the United Nations

On paper, the United Nations appears to have solved the problem of operational-level coordination, but the reality is quite different. In January 1999, the Secretary-General appointed Sergio Vieira de Mello as Under Secretary-General for Humanitarian Affairs, heading a new Office for the Coordination of Humanitarian Affairs (OCHA). De Mello is simultaneously the Emergency Relief Coordinator (ERC) who heads an Inter-Agency Standing Committee (IASC), chartered to coordinate efforts of all members of the UN family of organizations. At the operational level, a humanitarian coordinator would ensure coordination among all UN organizations. But some of these organizations resisted efforts by a predecessor, the Department of Humanitarian Affairs (DHA), to effect coordination. It remains to be seen whether OCHA will have more success than DHA did.

Within the U.S. government, the interagency process can be difficult, even though all agencies are ultimately subordinate to the President. Within the UN family of organizations, the interagency process is

---

Terms, Joint Publication 1-02; and service capstone documents, such as *Basic Aerospace Doctrine of the United States Air Force*, Air Force Manual 1-1. At the strategic level, civilian and military leaders define military goals necessary to achieve political purposes. At the operational level, senior military commanders employ military forces throughout a theater or area of operations. At the tactical level, unit commanders fight battles or accomplish those tasks associated with collateral missions such as humanitarian assistance.

[5]Frederick M. Burkle, Jr., Director of the Center of Excellence, sketched a table of this kind to illustrate that civilian agencies, excepting ICRC, lack operational-level control arrangements.

Table 8.1

Strategic, Operational, and Tactical Level Structures

Level	United States Government	United States Military	United Nations Family of Organizations	International Committee of the Red Cross	Non-government Organizations
Strategic	President, National Security Council, Principals Committee	National Command Authority; Chairman, Joint Chiefs of Staff; Joint Staff	Security Council, Inter-Agency Standing Committee, Office for the Coordination of Humanitarian Affairs (OCHA)	Council of Delegates, International Committee of the Red Cross (ICRC)	National and multinational headquarters
Operational	Special envoy; ambassador; commander-in-chief, unified command; commander, joint task force	Commander-in-chief, unified command; commander, joint task force	OCHA (humanitarian coordinator)? Lead agency? Regional Coordinator?	Delegates General	Ad hoc meetings? Civil-military Operations Center (CMOC)?
Tactical	Representatives of U.S. agencies; commanders of military units	Commanders of military units	Efforts of UN programs, funds, and specialized agencies	Efforts of ICRC and national societies	Efforts in the region or country

NOTE: Some titles and organizations are listed under multiple headings (e.g., the unified commands play an operational role in both the U.S. government and as part of the U.S. military) to reflect the multiple arenas in which they operate. A question mark suggests that the body identified makes a questionable contribution at the level indicated.

inherently more difficult because specialized agencies are not subordinate to the Secretary-General and therefore not compelled to coordinate, either at the strategic level through the ERC or at the operational level through a humanitarian coordinator. Moreover, in recent years a rival concept has emerged. During the protracted Bosnia conflict and more recently during the Kosovo crisis, the UNHCR has played the role of lead agency within the UN family. Such a role was natural because massive flows of refugees dominated in both cases and played to the UNHCR's specialty, but this de facto role supplants or disrupts the United Nations' formally declared arrangements. The danger is that a lead agency will give priority to its own requirements at the expense of an overall effort.

U.S. government officials approve the concept embodied by OCHA and provide funding for the OCHA-administered ReliefWeb. But they take a more reserved attitude toward the Military and Civil Defense Unit (MCDU) located in Geneva. MCDU is intended to ensure the effective use of military and civil defense assets, but it suffers from lack of support among those countries that provide the bulk of such assets during emergencies. Commonly, the United States declines to provide MCDU with data on available assets or to respond directly to requests for assets. MCDU is underfunded and will suffer from the recent ruling that prohibits member states in the United Nations from seconding military officers to the UN without charge.

UN organizations have limitations that can detract from their usefulness. Their coordination with the U.S. government through the U.S. Mission to the United Nations is uneven. They regularly meet with NGOs without inviting U.S. government participants and frequently ignore U.S. government requests for information. UNHCR and WFP are more nimble than other UN organizations, but even they can be slow and bureaucratic, particularly when compared with NGOs. By definition, UN organizations are responsible to member states, even when these states may be aiding combatants or otherwise contributing to a humanitarian crisis. In the interests of transparency, UN organizations may share information with such states, even to the detriment of military operations.

In contrast to NGOs, UN agencies work primarily with host governments, not directly with populations. As a result, they may focus on obtaining government approval rather than on working with local populations. This focus can distort relief efforts when host governments are repressive, corrupt, or incompetent. To maintain a good relationship with the host government, UN organizations may serve particular groups in favor rather than distribute aid according to need. In addition, the host government may misappropriate or profit from relief supplies.

Coordination Within the Red Cross and Red Crescent Movement

In addition to its other responsibilities, the ICRC directs and coordinates the actions of all components of the Red Cross and Red Crescent Movement. The ICRC does not direct operations from its headquarters in Geneva, relying instead on key individuals in the field, usually designated as Delegates General.

In the course of its duties, the ICRC acquires current information on topics of interest to the military. It will willingly share information concerning human needs, but it will not share information about armed forces. The ICRC learns much about armed forces simply because it is in nearly constant contact with them. Indeed, the ICRC maintains contacts with most of the armed groups in the world, including several that the U.S. government classifies as terrorist. But to preserve its neutrality and impartiality, the ICRC refuses on principle to collect or reveal any information about armed forces that would have intelligence value to an opponent. It will, however, provide information to military authorities and attend military briefings that deal with these aspects of a crisis.

The ICRC is eager to cooperate with the military on common humanitarian goals, but cooperation becomes difficult when the military is pursuing political goals that would compromise the ICRC's neutrality. For example, the ICRC cooperated closely with the U.S. military in Somalia prior to the intervention in December 1992. At the peak, the United States put six C-130 transport aircraft at the disposal of ICRC to conduct humanitarian flights into Somalia. After the United States intervened militarily, cooperation became more difficult and it ceased when the United States abandoned neutrality in its pursuit of the Somali warlord Mohammed Farah Aideed.

In recent years, the ICRC has increasingly encountered situations so chaotic that its neutrality and impartiality afford little protection. In Somalia, the ICRC found itself compelled to hire local guards. To maintain impartiality, these guards were drawn from all 31 warring clans and included people who would have looted ICRC supplies had they not been hired to guard them. Broadly speaking, the ICRC welcomes military action that provides general security, but it cannot accept military escort across lines of confrontation because belligerents would regard such escort as evidence that the ICRC was no longer neutral.

The ICRC's attitude toward the military is still evolving. After recently losing personnel in Chechnya, Sierra Leone, and other war zones, the ICRC has become painfully aware of the need for security. Moreover, it increasingly recognizes that it is no longer impartial when the aid it provides is diverted to combatants and warlords. In the past, ICRC delegates needed a direct order from Geneva to even converse with the military, much less cooperate with them, but today delegates have far more discretionary power.[6] The ICRC now sends its personnel to attend military exercises in an attempt to improve its cooperation with Western military forces.

Although the ICRC's zealous commitment to impartiality is frustrating at times for U.S. officials, respecting this commitment is vital for overall U.S. interests, particularly those of the military. The ICRC's impartiality enables it to visit U.S. prisoners of war. In Iraq and Somalia, the ICRC visited downed U.S. pilots, checking their status and demanding that their treatment comply with international conventions.[7]

Coordination Among NGOs

NGOs have no formal arrangements to promote coordination at the operational level, either within a single NGO or across all NGOs. At the strategic level, they have headquarters that generally advocate humanitarian action, raise funds for the organization, and ensure adherence to standards. At the tactical level, they have field offices that have day-to-day responsibility for programs. There is no intermediate-level arrangement to promote coordination until NGO representatives from different organizations meet to discuss a particular crisis, either in ad hoc meetings or in a setting such as a Civil-Military Operations Center (CMOC). Indeed, the CMOC—the operational body that facilitates NGO-military cooperation in the field—was designed to fill the operational void. All interested parties, including agencies of the United Nations, U.S. government agencies,

---

[6]Natsios (1995), p. 74.

[7]Bowden (1999), pp. 318–320.

NGOs, and local authorities should meet in the CMOC, which greatly facilitates cooperation.

Although NGOs appear anarchic, they have informal webs that promote coordination, at least among NGOs funded by a strong donor. For example, USAID expects that U.S.-funded NGOs will consult among themselves to develop practical divisions of labor. During crises, certain well-established, U.S.-based NGOs traditionally receive substantial funding from the U.S. government to provide immediate aid. These NGOs cooperate with each other to ensure that at least the overall U.S. effort is somewhat coherent. Among these NGOs are large organizations such as CARE, Catholic Relief Services (CRS), Save the Children (U.S. chapter), and World Vision.

Several individual NGOs often try to take the initiative to coordinate their fellow NGOs and plan for future developments. Although this coordination is usually ad hoc, it does allow for an effective response when the crisis in question develops slowly or is of limited scale. NGOs are particularly likely to take such initiative when operating in a highly dangerous area.

Some larger NGOs have central headquarters to promote coordination among their nationally based affiliates. For example, Adventist Development and Relief Agency (ADRA) has a headquarters in Silver Springs, Maryland, that oversees activities of ADRA worldwide organized under regional offices. CARE, Caritas, Concern, Doctors Without Borders, Mercy Corps International (MCI), Oxfam, Save the Children, and World Vision all have headquarters that coordinate efforts of the nationally based organizations.

In addition, many NGOs are members of professional organizations that promote professional standards. Examples include the U.S.-based InterAction, the European-based Voluntary Organizations in Cooperation in Emergency (VOICE), and the International Council of Voluntary Organizations (ICVA). InterAction is a membership organization of approximately 150 U.S.-based NGOs that forms standing committees and task forces to conduct projects on matters of mutual concern to its members. For example, the Sphere Project produced and disseminated a set of minimum standards for disaster response in such areas as water supply, sanitation, nutrition, food aid, shelter, and health services.   InterAction also provides a clearinghouse for the exchange of information and has descriptions of participating NGO activities in various countries.

COORDINATION STRUCTURES

Coordination structures vary from one operation to another, depending upon the situation, the mission, and the policies of host countries and donors.   There are four broad possibilities: host country lead, United Nations lead, alliance or coalition lead, and lead country.   These are not mutually exclusive alternatives and can be mingled during an operation.   The coordination structure shapes the operation, including coordination among actors, tasks to be performed, and rules of engagement.   The structures are supported at the local level by the CMOC.

Host Country Lead

When a host country's government is unimpaired, it will usually assert its sovereign right to authorize humanitarian relief as it sees fit. During natural disasters, a host country typically adopts an inclusive policy: It welcomes all the help it can get.   But during man-made disasters, a host country may curtail assistance that runs counter to its political goals.   For example, the Tutsi-dominated Rwandan government generally accepted humanitarian assistance during Operation Support Hope in 1994, but expelled 38 NGOs in December 1995 because they refused to accept direction.[8]   In some cases, the government may even have collapsed, causing near anarchy.   During relief operations in Somalia and Liberia, for example, there was no widely accepted central government that could take the lead.

Figure 8.2 is a simplified depiction of relationships during disaster relief following Hurricane Mitch, which struck the Caribbean and Central America in October 1998.   Each affected country had direct working relationships with international organizations, the Pan American Health Organization, the ICRC, and NGOs.   In each country, the U.S. ambassador or chargé d'affaires declared a disaster,

---

[8]Action Against Hunger (1999), p. 28.

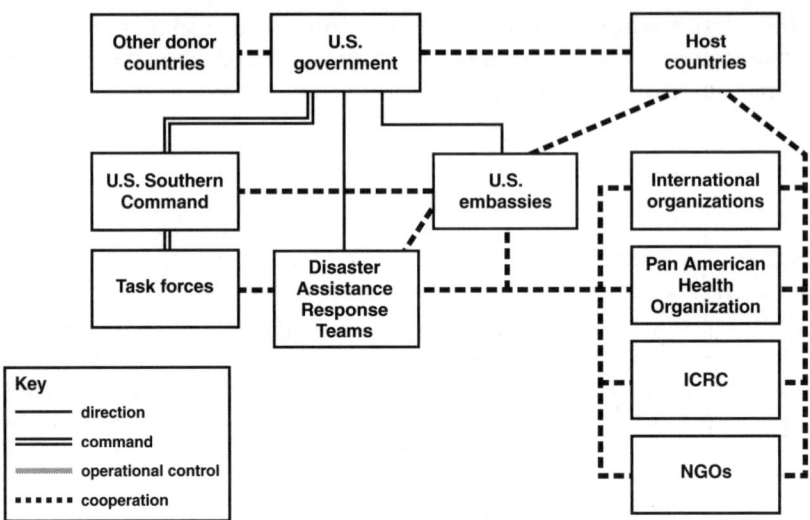

NOTE: Direction: management, control *(Webster's Unabridged Dictionary,* 1979); command: the authority a commander in the Armed Forces lawfully exercises over a subordinate *(Department of Defense Dictionary of Military and Associated Terms,* 23 March 1994, amended through 6 April 1999, Joint Pub 1-02); operational control: transferable command authority exercised below the level of combatant command (Joint Pub 1-02); cooperation: working or operating together to one end *(Webster's).*

Figure 8.2—Host Country Lead

making that country eligible for emergency assistance from the United States. OFDA sent Disaster Assistance Response Teams (DARTs) to assess the situation and help coordinate the U.S. response. The U.S. Southern Command (USSOUTHCOM) formed task forces that coordinated with the DARTs and were responsive to country teams in the U.S. embassies that were in contact with host country governments. NGOs and the relief community cooperated, but no directed activity occurred even though the U.S. government was leading the relief effort.

United Nations Lead

When a host country's government is impaired, but outside powers do not intervene decisively, agencies of the United Nations may assume coordinating roles. Within the UN family of organizations, there are two broad possibilities: coordination through the Emergency Relief Coordinator (ERC) or through a lead agency, most likely the UNHCR.

Figure 8.3 offers a simplified view of relationships among agencies supporting the humanitarian effort in Bosnia prior to the Dayton Agreement. During this period, the United States airlifted supplies into Sarajevo and airdropped supplies into Muslim-held enclaves in concert with its NATO allies. According to formal procedures, the ERC, working through the Inter-Agency Standing Committee (IASC) and OCHA, "will mobilize and coordinate collective efforts of the international community, in particular those of the UN system."[9] But the United Nations has continually failed to implement this model. In several recent crises, UNHCR has acted as a lead agency—

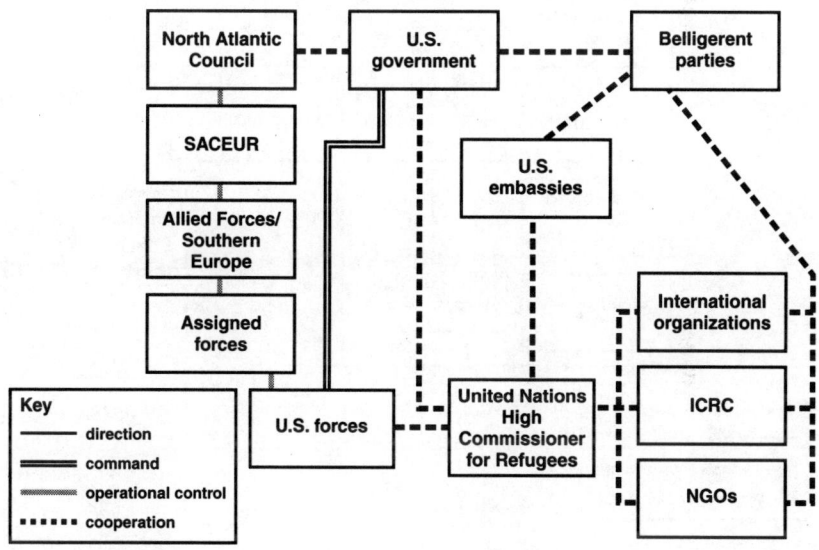

Figure 8.3—United Nations Lead

---

[9]General Assembly Resolution 46/182, which created the predecessor organization Department for Humanitarian Affairs (DHA).

for example in Bosnia prior to the Dayton Agreements and currently in Kosovo.[10]

Alliance or Coalition Lead

During a humanitarian crisis caused by conflict, an alliance or coalition of willing powers, often identical with the major donors, might coordinate assistance. Assistance to Bosnia subsequent to the Dayton Agreement followed this pattern.

Figure 8.4 presents a simplified picture of relationships after Dayton. The highly complex post-Dayton arrangements include roles for the

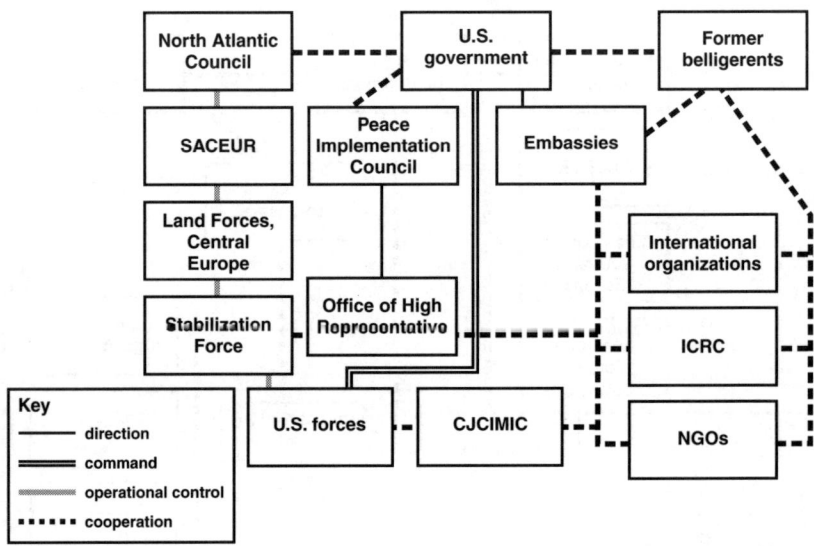

Figure 8.4—Alliance or Coalition Lead

---

[10]Although the UNHCR retained formal coordinating responsibility for relief efforts in and around Kosovo [now under the United Nations Interim Administration in Kosovo (UNMIK)], the coordinating and relief management functions of UNHCR proved inadequate to the task and "migrated" in practice to NATO. Even in Bosnia, the role of the UNHCR was to some extent overshadowed by NATO and OSCE activities and, at the political level, by the role of the Contact Group.

United Nations and other IOs, a Peace Implementation Council (PIC), the North Atlantic Treaty Organization (NATO), OSCE, and, of course, the former belligerents. Such complex arrangements are workable because the same powers are present in all these organizations and they coordinate among themselves at the policy level through the Contact Group and other means. These same powers (plus Japan) are also the major donors of humanitarian aid. Acting as leader of this alliance, the United States helps to organize donors' conferences under the auspices of the World Bank, which publishes and oversees an overall plan for the reconstruction of Bosnia. NATO forces coordinate with civilian agencies through Combined Joint Civil Military Cooperation (CJCIMIC)—active and reserve civil affairs personnel from around the world who support the Office of the High Representative (OHR) and serve as a link between military and civilian agencies.

Although NATO is the most effective regional alliance, others might also take the lead. In Liberia, the Economic Community of West African States (ECOWAS) took the lead in forming an intervention force. In Africa, the United States may work with non-NATO regional alliances led by important African states, such as Nigeria or South Africa.

Lead Country

One country may take the lead and invite other countries to join it. In this simplest case, the lead country assumes a responsibility for coordination. For example, the United States was lead country during operations Provide Comfort I in Iraq (April–July 1991) and Restore Hope in Somalia (December 1992–May 1993). Other major powers may play this role, as has France in sub-Saharan Africa.

Provide Comfort I was a humanitarian operation to ensure survival of Kurds who had fled from Saddam Hussein's forces in early April 1991 following the Persian Gulf War. Some 750,000 refugees were at risk from exposure, thirst, hunger, and disease, and at peak some 1500 were dying each day. On April 5, the United Nations Security Council passed Resolution 688 authorizing use of force to protect relief operations for these refugees. Under this resolution, the United States organized a joint task force, soon expanded to a combined task force, to secure areas of northern Iraq, deliver emergency supplies, and assist return of the refugees to their homes. Eleven other nations provided military forces and all (except German forces) were eventually controlled by Combined Task Force Provide Comfort[11] commanded by Lt. Gen. (USA) John M. Shalikashvili. OFDA deployed two DARTs to Turkey to help link civilian and military efforts.

Figure 8.5 shows key relationships during Operation Restore Hope. Restore Hope was intended to ensure survival of Somalis threatened by starvation and disease as a result of interminable violence among rival clans. At peak during 1992, some 1,500,000 Somalis were at risk and some 300,000 are estimated to have died. After a small U.N.-controlled operation proved ineffective, the United States offered to lead a larger military force. On the basis of this offer, the UN Security Council passed Resolution 794 authorizing the use of force to establish a secure environment for relief operations. Several other

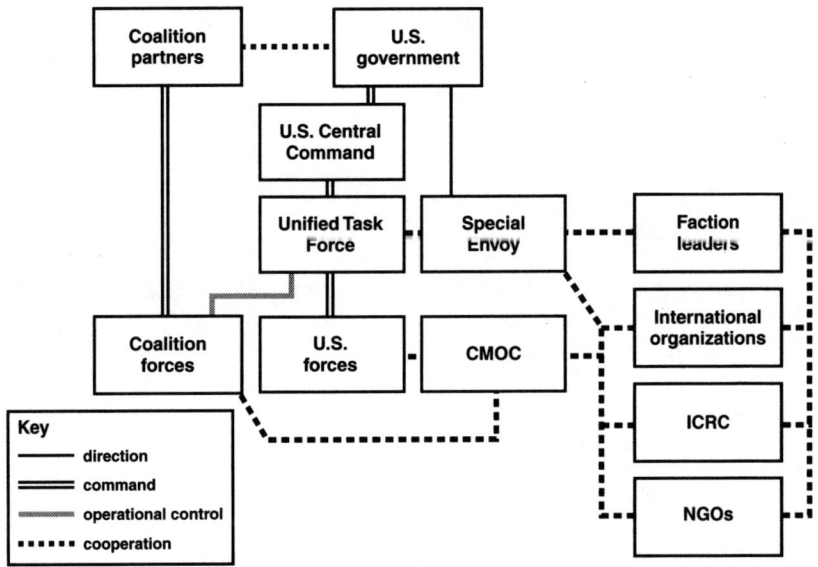

Figure 8.5—Lead Country

---

[11]A second task force, designated Task Force Encourage Hope, was formed to construct resettlement camps.

countries also deployed forces to Somalia in anticipation of a larger UN-controlled operation to follow. Most of these forces were temporarily controlled by the Unified Task Force (UNITAF) led by Lt. Gen. (USMC) Robert Johnson, commander of I Marine Expeditionary Force. The United States sent Ambassador Robert B. Oakley as a special envoy to coordinate all U.S. civilian activities in Somalia, provide political advice to Johnson, and work closely with NGOs.[12]

UNITAF coordinated with international and nongovernment organizations through a CMOC. There was a central CMOC in Mogadishu and a satellite center in each of eight Humanitarian Relief Sectors (Baidoa, Baledogle, Bardera, Belet, Gianlalassi, Kismayo, Oddur, and Uen). UNITAF took responsibility for airport and seaport operations. It provided security to aid convoys and to air distribution points, and it also dismantled unauthorized checkpoints and enforced an increasingly stringent weapons control policy.

Limits to Coordination Structures

Although the above coordination structures provide some organization to a relief effort, cooperation may still be limited or imperfect. The structures discussed above reflect what has been done on an ad hoc basis. Because the structures often vary considerably from crisis to crisis, establishing relationships and procedures is difficult. Furthermore, the structures rely on NGOs to coordinate their activities but do not direct their effort in any way. Finally, the structures are often highly complex, with many actors and uncertain control and coordination arrangements.

CENTER OF EXCELLENCE

To improve NGO-military familiarity and coordination, the Department of Defense currently sponsors the Center of Excellence (COE) in Disaster Management & Humanitarian Assistance, located in Hawaii and affiliated with U.S. Pacific Command (USPACOM).[13]

---

[12]Hirsch and Oakley (1995), p.50.

[13]As of this writing, a COE is being established that will be affiliated with U.S. Southern Command.

The COE is a unique organization that focuses on improving coordination at the operational level. The COE builds on the experiences of previous operations to improve civilian and military response.[14]

COE develops training materials and presents courses in humanitarian assistance to both military and humanitarian audiences. Courses include the Combined Humanitarian Assistance Response Training (CHART) and Health Emergencies in Large Populations (HELP). COE developed CHART to introduce civilian and military participants to the fundamentals of relief operations. HELP is a longer, more specialized course originally developed by ICRC. Under current procedures, COE conducts these courses without cost at sites specified by clients.

COE provides support to training, games, and exercises conducted by the military, such as Brave Knight, Prairie Warrior, and Emerald Express. It identifies appropriate subject-matter experts, assists in development of scenarios, plays roles, and assesses relief strategies. COE facilitates flows of information among international organizations, NGOs, government agencies, and the military through its Virtual Information Center and the Pacific Disaster Management Network (PDMIN). COE is currently developing the Combined Event Notification Technology and Unified Reporting (CENTAUR), specialized software originally sponsored by the United Nations Children's Fund (UNICEF). COE expected to begin field testing CENTAUR in 1999 and hopes to persuade not only UNICEF but also other UN organizations to adopt the system. The fundamental problem may be

---

[14]Congressional mandate established the COE in October 1994. Senator Daniel K. Inouye, Democrat from Hawaii, then a senior member of the Appropriations Committee, was the congressional sponsor. Dr. Frederick M. Burkle, Jr., Chairman of the Division of Emergency Medicine, University of Hawaii Schools of Medicine and Public Health, promoted the concept. He envisioned an organization that would help draw together disparate agencies involved in humanitarian assistance. COE currently operates under draft Articles of Association that define an Advisory Committee that includes the sponsoring U.S. Senator; Commander-in-Chief, USPACOM (USCINCPAC); Commanding General, Tripler Army Medical Center (Tripler AMC); President, University of Hawaii; and the Director, COE. Reflecting its origins, COE initially tended to have its closest relationship with Tripler AMC, but in recent years it has begun to develop closer relations with USPACOM. COE currently has 26 personnel, many seconded from other organizations including the Center for Disease Control (CDC), and an annual budget of $5 million. Center of Excellence in Disaster Management & Humanitarian Assistance (1998), p. 7.

to persuade these organizations to share information fully. COE also sponsors research projects on topics that cut across organizational lines, such as development of measures of effectiveness for health in refugee camps.

Beyond these activities, COE provides a source of expertise in humanitarian assistance that is constantly available to USPACOM. COE personnel are broadly familiar with every aspect of humanitarian assistance and are personally acquainted with patterns of need and the assets available to address these needs through the PACOM area of responsibility (AOR). Therefore, personnel drawn from COE would be well qualified to fulfill the role of humanitarian advisor to the Commander-in-Chief, U.S. Pacific Command (CINCPAC).

Despite its many advantages, military coordination with potential partners in a humanitarian crisis is often difficult because there is no official structure to coordinate activities. Particularly at the operational level, coordination among NGOs, IOs, donor governments, and military forces lacks structure. The structures described above, including the COE, offer only a limited means of coordinating a relief effort. In addition, as discussed further in the following chapter, many relief agencies have characteristics that hinder coordination and may make them difficult partners.

Chapter Nine

BARRIERS TO IMPROVED COORDINATION WITH RELIEF AGENCIES

Coordination between the military and relief partners, particularly NGOs, is often uneven and uncertain. NGOs can be difficult partners, especially for the military. There is a wide gap in organizational culture, and NGOs are inhibited by their concern for neutrality and impartiality. NGOs also do not plan well, making cooperation before a crisis difficult. There is an evident lack of mutual familiarity, and NGOs are often reluctant to share information with the military. NGOs and the military may compete for publicity and they have different time horizons. Finally, NGOs are not certain of the military's true commitment to humanitarian missions.

The barriers to better military-NGO coordination are numerous but not insurmountable. Indeed, during major operations, strongly motivated people in both camps usually find ways to surmount these barriers, but valuable time is lost inventing and reinventing these solutions. Relationships have improved in recent years, but considerable progress is necessary before both sides can realize the advantages of improved cooperation. This chapter describes common barriers and notes progress in reducing them.

DISPARATE ORGANIZATIONAL CULTURES

Differences among organizational cultures are a formidable barrier to NGO-military coordination. Differences include:

- *Hierarchies versus decentralization.* NGO organizational structure is very different from that of the military. Most NGOs are managed in a highly decentralized manner, with scope for initiative in the field. Typically, they prefer to work by consensus rather than responding to direction. Rather than being hierarchical, with a clear and orderly assignment of responsibility and authority, NGO structure is usually egalitarian, with much debate required before a consensus-based decision is reached. Accustomed to this autonomy, many NGO personnel have little patience with military hierarchies. They tend to resent military officers' typical question: "who's in charge?"

- *Discomfort with the use of force.* Some NGO personnel are skeptical of the morality and efficacy of military force. They are accustomed to regarding the military as part of the problem and remain critical of the military even while it provides essential support.[1] At times this discomfort reflects an overall unease about military operations, which can interfere with information sharing. This can be accentuated when the meetings are held on a military facility and NGOs are required to submit to elaborate checkpoint procedures before entering.[2]

- *Different ways of life.* The values and lifestyles of many NGO employees are not always compatible with values prevalent in the military. The NGO community features respectable church-based aid multinationals represented by nuns and sophisticated groups of highly qualified scientific, technical, and medical professionals, but it also includes "a colorful collection of Woodstock grads, former Merry Pranksters and other assorted acid-heads, eco-freaks, save-the-whalers, doomsday mystics,

---

[1]Some NGO personnel can be abusive to the military even as they seek military assistance. In Somalia, for example, NGOs demanded transportation, security, and communications assistance yet wanted the military to minimize its presence. Their attitude was described by one NGO observer as: "Give us a ride. Save our lives. But don't come near us."

[2]For example, when NGO representatives met with U.S. military staff in Tuzla, they were intimidated by the security precautions, even though the military treated them with deference. The CMOC was located inside Task Force Eagle's headquarters facilities, forcing the relief agencies to go through security at the base perimeter (as well as to travel several miles to attend the meetings). Subsequently, many of these NGO representatives avoided interaction with the military.

poets and hangers-on."[3] Some NGO personnel are amateurish, have strange personal biographies, or come from countries hostile to the United States.

- *Skepticism about force protection.* NGOs often wonder why well-armed military units emphasize force protection while working in areas where NGOs have long operated without protection. In addition, NGO personnel can be intimidated by displays of military force.

- *Secrecy.* NGOs are highly transparent organizations. They usually publicize their operations to attract funding from international, governmental, and private donors. As a result, they have little understanding for military secrecy and tend to resent the classification system.

Because of these cultural differences, NGO and military officials may not understand each other's priorities or procedures and resent what they see as indifference on the other side.

These differences, however, may be overstated and mask similarities that make coordination easier. Like the military, NGO personnel are often highly idealistic and willing to dedicate their lives to helping others. Many NGO personnel are exceptionally brave, living and working in war zones where banditry and disease are common. NGO personnel, especially those in the field, are focused on the mission and willing to use work-arounds or otherwise deviate from accepted procedures to finish the job. Finally, like the military, NGO personnel are comfortable with foreign cultures and ideas and have an international perspective.

CONCERNS ABOUT NEUTRALITY AND IMPARTIALITY

NGOs rely heavily on their neutrality to protect themselves.[4] They seek to project a certain image: They want local authorities and

---

[3]Rowland (1973), p. 1.

[4]The "Code of Conduct for NGOs in Disaster Relief" spearheaded by the ICRC, the Red Crescent, Save the Children, Oxfam, the Lutheran World Federation, and the World Council of Churches lists the most important principles that should guide disaster response NGOs. To the point of redundancy, fully the first four of these principles reit-

warring parties to feel that NGO personnel are basically harmless, possibly even useful, while attacking them would needlessly bring bad press, anger in the countries they are nationals of, future boycotts by their organization when their side is the one that needs help, and so on. This explains the NGOs' sometimes baffling attitude toward military protection. Even though they may need an armed guard or a military escort in a particular situation, they may fear that, in the long run, association with the military threatens their image and endangers them. As Jean-François Vidal of Action Against Hunger noted:

> Our protection is usually the perception people have of us. We are endangered when we appear close to the military. We have no limits on sharing humanitarian information with the military. Reporting incidents is not a problem. But sharing military intelligence, such as strength and weaponry of belligerents, is dangerous for us. The farther we are from the guns, the better we feel.[5]

In the field, NGO operatives often walk a fine line. By barter, by compromise, by charm, or by mobilizing public opinion, they try to overcome obstacles as they arise. This can mean disregarding or deliberately flouting the distinction between friend and foe. As John Ashton of Response International noted in an interview:

> When the UN closes the line, that doesn't mean we stop. And people respect that. You have to establish relationships, find out what people want. We would talk to the Serb soldiers and they would say, my uncle needs this kind of medication, my niece needs that, my brother needs this, etc. We would get them the stuff, and in exchange they allowed things to go into Sarajevo. Everybody has needs, even the aggressor. Of course they use aid as a leverage point but they can be flexible once they trust you.

In essence, these organizations stay safe by making themselves nonthreatening: Their weakness protects them. The ICRC and many NGOs as well also embrace neutrality in their mission. They seek to

---

erate the goals of independence and autonomy, emphasizing how fundamental these values are to NGOs.

[5]Authors' interview.

provide aid to all individuals, regardless of their political position or past activities.

Preserving neutrality and impartiality, however, becomes difficult—and often impossible—when the United Nations or a member state such as the United States undertakes enforcement. As Joelle Tanguy, the Executive Director of MSF, noted:

> I'm afraid that in the minds of Americans and Europeans, the military and the relief organizations are working on one side of the war together. . . . We're all part of the same operation, but we can't be. Independence is our main asset—to be able to walk into a war zone and act as independent relief workers.[6]

In Somalia, for example, the United States and UNOSOM II (the second UN Operation in Somalia) attempted to apprehend the Somali warlord Aideed, thereby forfeiting impartiality, at least in the eyes of his supporters.[7] NGOs feared that this loss of neutrality would impede their operations and lead belligerents to see them as allied with combatants, and they worried that a military conflict could lead to their personnel being targeted. World Vision personnel were, in fact, attacked by militia forces expressing their displeasure with the United States-led enforcement. Similarly, even before the NATO bombing campaign in Kosovo, some NGOs avoided ties to the military, in part because many of their third-country national employees were hostile to NATO. Once the bombing began, impartiality became far harder.

Because the United States is viewed as having a global agenda, NGOs may fear being seen as a pawn in U.S. policy even in cases like

---

[6]Becker (1999).

[7]The concepts of neutrality and impartiality are not always well understood or correctly applied. Neutrality implies that all parties will be equally affected by an action. But no peace operation, not even unarmed monitoring, will be likely to affect all parties equally and therefore none is neutral. Impartiality implies that the United Nations, normally the Security Council, believes all parties share responsibility and therefore refuses to identify aggressor or victim. Peace operations are or should be impartial. In Somalia, the Security Council was impartial in the sense that it would presumably have attempted to enforce the peace agreements on any party found in violation of them—particularly if, as Aideed did, they ambushed UN peacekeepers. But even Western commentators failed to understand this distinction, and Aideed and his supporters believed anyway that they were being unfairly singled out.

Rwanda, where the United States concern was almost entirely humanitarian. ICRC officials have more difficulty working with the U.S. military than with those of smaller powers, such as Canada or Sweden, because the United States usually has a political agenda—or is seen as having one.[8] NGOs thus often guard against even the appearance of partiality by avoiding unnecessary contact with military staff. As one NGO official noted, "walking into a bar with an officer can hurt our impartiality." Antoine Gerard of MSF noted in an interview:

> We try as much as we can to differentiate from any military that is present. The image of cooperating with the air force is scary for us. This would mean recognizing that we are part of the conflict, and it would send a confusing message to the populations we are trying to help.

This concern hinders closer personal relations and the communication that can ensure smooth operations.

NGOs themselves, however, often have trouble living up to their ideals of neutrality. Neutrality and the aim of remaining extraneous to a conflict are often unrealistic goals, perhaps particularly in contemporary conflicts. NGOs are aware of this and engage in considerable soul-searching. In a typical position paper on this issue, prepared by and for NGOs, Hugo Slim notes that:

> in any analysis of the causes of violent conflict, it is very important to recognize the part NGOs and aid can play in escalating conflict. Any analysis of violence should recognize how complicated responsible emergency work is during conflict and how NGO programs can so easily become part of the cycle of violence.[9]

Similarly, an analysis of NGO work in Mozambique and Sudan notes that NGOs may contribute to the fighting inadvertently, because their relief is a valued commodity by locals, which makes them a target for rival militias. Indeed, the presence of NGOs can even contribute to the suffering of innocents: Unscrupulous warlords may

---

[8]Seiple (1996), p. 45; interviews with relief officials corroborate this point.
[9]Slim (1996).

increase overall suffering and destitution to attract relief they can control and parcel out for their own supporters.[10] NGOs also at times ignore the human rights problems their aid inadvertently abets. NGOs remained in Zaire and treated Hutu refugees from Rwanda, even though their assistance directly aided Hutu warlords who had committed a genocide in Rwanda and were continuing cross-border raids.

The situation becomes stickier still in active-combat situations. NGOs are not above purchasing access, safe passage, or permits with bribes. They thus strengthen the warlords who cause much of the suffering.[11] Currently, in Afghanistan, the usually fastidious MSF has broken ranks with other NGOs by providing money and support for the Taliban and letting them dictate the terms of medical treatment, in order to be allowed to remain.[12]

LIMITED NGO ABILITY TO PLAN

NGOs are often accused of being chaotic and uncoordinated in their activities. Although NGOs want to improve planning—and at times they have coordinated their actions impressively—they face objective limits to how well they *can* plan.

The NGO emphasis on impartiality and independence hinders long-term planning with the military. Cooperation that requires a formal, public relationship, or seems to limit the autonomy of NGOs, will probably be resisted by NGO leaders. This independence is an asset that allows NGOs to operate where organizations tied to the U.S.

---

[10]Keen and Wilson (1994).

[11]Whether the chance to help the victims justifies the compromised principles can be a difficult call. German Greens were ridiculed when, following their visit to the Bosnian war zone, they refused to give their bulletproof vests to Bosnian civilians who requested them, on the grounds that this would amount to supplying one side over another with war-related items.

[12]The arrest, in April 1999, of two Australian CARE humanitarian aid workers, and the announced intent of the Milosevic government to put them on trial as NATO spies, represents a new and alarming watershed. In their information exchanges with the military, and precisely to avoid charges such as these, NGOs officially aim to impart only facts relevant to the humanitarian crisis and nothing of military use. Incidents such as this may inspire the NGOs to seek greater distance from the military or it may drive them closer to whatever protection the military can provide.

government are not welcome, but it hinders coordination beyond ad hoc measures.

In addition to concerns about autonomy, many analysts suggest that poor NGO planning arises from the nature of the problems being addressed: Emergencies, they point out, are by definition unexpected, abrupt, and unpredictable events and are thus resistant to structure and preplanning. Essential goods are often missing, unavailable, or delayed. A generator may be en route, but the airport is not functioning; it may have arrived but cannot be unloaded because the workers are not there; or it may have been unloaded but there is no secure storage or forward transportation; and so on. Information may be sketchy and not always reliable. An NGO may have to deal with the national police force and the official military, one or more rival militias, peacekeeping troops, international agencies, representatives of various governments and of different militaries, the media, and other NGOs, all of which have different agendas, infrastructures, and rules.

The nature of relief work produces a frustrating and at times fatal combination of redundancy and gaps. Information flows may be poor, particularly early in a crisis. There have literally been cases, in African famines, where camps received boxes of eating utensils but not any food. One location may receive the vaccines and another, hundreds of miles away, the syringes for dispensing them. Lack of information exacerbates the problems, since workers on the ground cannot be sure if or when urgently needed supplies will arrive.[13]

The "chaos argument," while having some validity, should not be overstated. The argument that the NGOs' chaotic operating environment produces poor planning is shaky; the same is true of wars, which have produced institutions, such as the military staff,

---

[13]Balancing the massive emergency-care needs against the danger of an epidemic, medical workers in Sudan reluctantly decided they could no longer wait for the vaccination guns that would have allowed a rapid and efficient inoculation, and instead they began vaccinating by syringe. Given the small number of aid workers in this medical project and the large number of refugees, this meant neglecting other essential operations, such as the infant oral rehydration program and critical care. Neglecting these meant that people would die, but, given the poor hygiene conditions and unsafe water supply, the danger of an epidemic seemed more grave. Two days later the guns arrived.

that are the very epitome of structure and preplanning. Many of the worst NGO problems result from inadequate coordination and a cumbersome start-up process. In contrast to the military, no NGO institution has responsibility for the entire effort. There is redundancy in some areas and complete failure in others. Too many people are on location without clear division of labor; the processing of each task consists of long sequences with many opportunities for things to go wrong or be delayed; and there is often no command structure or even anyone reliably in charge. Even if everyone involved has the same goal in mind and is of good will—a precondition that definitely does not hold true in most international emergencies—the involvement of so many people and agencies creates clumsiness and inefficiency. The NGOs' distrust of hierarchy hinders attempts to bring order to this chaos.

The sheer number of institutions, and the small size of many of them, can hinder coordination. Relief work requires the interplay of multiple actors and sovereignties, all of whom have different agendas, structures, and chains of command, and many of whom are in a state of rivalry or hostility with each other. In any given crisis, multiple levels of coordination are necessary with and between national governments, international organizations, national aid organizations, and NGOs. NGOs operate in an environment that is characterized by the absence of authority or, more often, the presence of several competing, sometimes even warring authorities.

NGO problems with planning can begin with the donors, who range from individuals filling up cardboard boxes with their family's outgrown winter clothes to church groups running collection drives to businesses and corporations of all sizes and compositions. These sponsors do not necessarily give what is needed; they give what they can spare and think appropriate, which can include medication well past its expiration date, clothing unsuitable to the climate, and funding tied to conditions that hamper the recipients. Clearly, it would be sensible to stockpile donations independently of a crisis, when one has the leisure to sort and review and catalogue, and certainly this happens too, but human psychology is such that the bulk of donations pour in when a crisis occurs and segments of the world public, for reasons of proximity, dramatic camera footage, or some other emotional affinity, urgently feel moved to help and give.

NGOs are also affected by constraints and traditions within their own community. For instance, many NGOs are accustomed to subsector coordination on the basis of some kind of affinity. Church organizations tend to coordinate with other church organizations, medical groups with other medical groups, and so on. These organizations may not talk to others outside their community.

Over time, many of these problems are sorted out. NGOs in the field establish structures for communicating and arranging a division of labor. Personal ties in the relief community are often strong, creating impressive networks that enable experienced individuals to informally coordinate their activities with others. In the early days of a crisis, however, the lack of advanced planning is particularly troublesome.

AMBIVALENCE ABOUT SHARING INFORMATION

Although NGOs are often open with information concerning the needs of suffering people, they may be reluctant to share other information with the U.S. military. NGOs are hesitant to provide information on personnel and staff, including third-country nationals. They are often particularly reluctant to share information on the host government, fearing that it will compromise their access to crisis zones.

Some NGO officials worry that the military seeks to collect information that goes well beyond the immediate crisis. Similarly, the ICRC fears being seen as spies—by both local parties and U.S. officials—because they regularly meet with people on all sides of a conflict.

NGOs do not want information-sharing to be a one-way street and resent what they deem as one-sided information exchanges. Military concerns about classification further hinder information-sharing. In Somalia, for example, many NGO members became frustrated by the military's refusal to discuss fighting that occurred in NGO areas of operation. For example, in the Civil-Military Operations Center NGO participants wondered, "What isn't the military telling us?" If the military is not up-front about what it is not sharing, such as information on the movement of forces, NGOs may believe they are hiding information as a matter of policy. As one relief official noted:

In Somalia, the military would open meetings with weather reports, but we all knew what the weather was and it seldom varied. Then an NGO would mention that fighting had occurred in its area during the night but the military would refuse to discuss the topic because it was classified. Thus the military communicated useless information but declined to share information that could have been helpful. We wanted to know whether the military was informed about the security situation and whether it intended to react to outbreaks of fighting. The military cannot expect NGOs to provide information unless it is also willing to talk.

NGOs regularly trade information among themselves and expect the military to trade as well.

The information NGOs provide is at times skewed. Relief personnel new to the crisis area may know little about local conditions or actors beyond their immediate area of operation. Relief agencies also have a financial interest in dramatizing a crisis: They know that day-to-day misery receives far less support than do sudden, heart-wrenching crises that grab media attention. Thus, they may play up suffering to gain funding for their less-glamorous activities.

As with other generalizations about NGOs, this problem varies from organization to organization. The larger, more-established NGOs are less likely to manipulate information or resist cooperation with the military, largely because they expect to work with the military again in the future. Smaller NGOs, and many non-U.S. NGOs, are often far more reluctant to share information with the military.

In general, NGOs are more willing to share information with elements of the U.S. government who are not in uniform. USAID personnel or civil affairs officers, for example, are considered more suitable for information exchanges, even though these officials then relay the information to the military. As with other NGO concerns, much of this distinction boils down to perception: A uniformed military officer is often more suspect than other individuals regardless of the nature of the mission or that individual's activities.

COMPETITION FOR PUBLICITY

Relief agencies compete against one another to gain scarce funds, a competition that hinders cooperation among them and with the U.S.

military. The more dramatic and heart-wrenching the story NGOs can tell to potential donors, the more money they are able to raise.[14] In practice, this may lead NGOs to devote considerable attention to public relations and the media, to prove to donors and the public at large that they are active.[15] Even UN agencies share this concern. As one WFP official noted, "It isn't just doing the good deed. We have to be seen doing it."[16]

In their drive for publicity, NGOs may seek a visible role in the relief effort even when their participation contributes relatively little. In the early days of a crisis, some NGOs show up to demonstrate to their donors that they are present and contributing—an image that makes it easier for them to secure funding. This visible presence, however, can interfere with the smooth flow of aid and personnel to a distressed region. Moreover, it may lead to the neglect of less-glamorous elements of an aid operation, such as sanitation. NGO competition with one another and the military often increases as a crisis matures. Early on, there are simply too few people and too many problems. Over time, however, NGOs begin to compete for missions, both among themselves and with the military.

Publicity concerns also contribute to inefficient resource allocation. During the April 1999 refugee crisis in Kosovo, experts explained on television why only cash donations made sense, while at the same time the Kosovar Relief Fund in New York and Washington was busily calling for donations of cases of bottled water, canned goods, and blankets. Fund officials were thrilled to have persuaded Mayor Guiliani to open New York fire stations to receive these goods, oblivious to the fact that everything would then have to be flown a significant distance at great expense. Such donation drives have the advantage of being tangible and visible, and thus perhaps carry a public relations benefit, but the opportunity cost is high. People who went to the trouble of dropping off bags of canned soup would almost certainly have been willing to donate cash instead but will now consider that they have done their bit.

---

[14]Natsios (1995), p. 71.

[15]Seiple (1996), p. 86.

[16]Pope (1999).

Fund-raising sensitivity also may cause inadvertent resentment of the military. Military forces quickly attract the camera. Thus, when the military is in the field, it often becomes harder for an NGO to claim credit for relief activities or otherwise raise money.

NGOs' desires to gain recognition for their efforts can contribute to political pressure on the military operation. NGOs—both local and national—will try to work through Congress to ensure that their contribution receives the priority they believe it deserves. If they deem it necessary, NGOs can generate a storm of controversy. This can lead to political decisions taking precedence over those of relief professionals.

VARYING TIME HORIZONS

Because they will be on the scene after the military departs, NGOs have a different perspective on relief operations. NGOs cannot afford poor relations with locals, no matter how thuggish. As one NGO official noted about Haiti, "NGOs were there before the military arrived and remained there afterwards." Thus, they must weigh the benefits of short-term cooperation with the military against the possible negative consequences of long-term alienation.

The different time horizon gives NGOs a different perspective on U.S. offers of security assistance. Although in the short term an NGO may be safer because of U.S. protection, the protection may fatally compromise the NGO in the eyes of the locals after the United States departs. Thus NGOs may be reluctant to accept U.S. offers of security if they plan to continue operations in the country over the long term. Moreover, NGO officials have learned from past experience that the U.S. military can depart quickly with little warning.

NGOs, particularly those involved in long-term development work, and the military often measure success differently.[17] Military officials may arrive on the scene of an intervention with quantitative

---

[17]NGOs, however, may ignore long-term needs. Donor countries often care little about long-term relief, focusing their attention on highly visible crises. As a result, there is less incentive for NGOs to emphasize long-term development. Similarly, the presence of the military often concentrates political attention on immediate gains. Forman and Parhad (1997).

measures of success, such as reducing mortality rates or restoring an infrastructure. For NGOs, success may be measured by using resources efficiently, not by solving the problem.[18]

NGOs are particularly skeptical of the military's focus on the "exit strategy"—a complaint almost universally shared by NGO interlocutors. Because NGOs will remain in the country after the military has departed, they do not share the military's focus on accomplishing the tasks at hand to facilitate an on-time departure. They may see this talk as proof that the military is not committed to solving the problem in a thorough way.

MUTUAL LACK OF FAMILIARITY

Although knowledge has grown in the last decade, military officers and NGO officials often have little understanding of each other's institutions and operating procedures. Many military officials lack an understanding of the distinct charters and doctrines of NGOs, failing to recognize that what works with the IRC will not work with the ICRC.[19] In turn, aid organizations criticize the military for not understanding their hierarchies. As one aid official noted in an interview, "The military should accord the heads of major NGOs the respect normally granted to a general officer."

The military may not be familiar with important NGOs in the AOR. Before IFOR (Implementation Force, Operation Joint Endeavor), the United States European Command (USEUCOM) was not aware of how to contact NGOs in the area. Similar problems occurred in operations in Somalia, Rwanda, and Haiti, where the NGOs were treated as an afterthought despite their important role in an operation.

The reason for this lack of knowledge is institutional. Although many officers have worked with relief agencies over the past decade, little effort has been made to retain this knowledge. In the military, only civil affairs officials routinely work with NGOs, and almost all these

---

[18]UNHCR (1995), p. 15.

[19]Dworken (1996), pp. 19–20.

capabilities are in the reserve forces.[20] Obtaining knowledge before a crisis, when reserve forces are less likely to be deployed, is therefore difficult. Although local country teams bear some responsibility for tracking NGO activities, in practice local embassies are often overextended and have little knowledge of aid agency activities. In the Air Force in particular, there is no institutional responsibility for tracking NGO activities and ensuring liaison with important NGOs.

Many NGO officials see little need to volunteer information on their activities.[21] In Rwanda, NGOs, the United Nations, and the U.S. military were all unaware of which NGOs were operating in the region.[22] Many NGOs do not register with the U.S. embassy or otherwise make their presence known. In Rwanda, Somalia, and other crises, NGOs often simply appeared without prior arrangements to be received.[23]

Ignorance of the military on the NGO side compounds the problem. NGO officials often are completely ignorant of the military. Military organization, hierarchies, and capabilities may be understood through movies rather than through experience. Even ICRC officials have little knowledge of the military or how it operates despite their regular presence in war zones. Discovering existing, well-established military programs for providing lift—such as Denton Program flights—often occurs by chance.

As a result of this ignorance, aid organizations may have unrealistic demands of what the military can provide. In Somalia, for example, aid organization personnel expected an almost instant deployment of U.S. personnel throughout Somalia after the decision to intervene

---

[20]Barnes (1989).

[21]Seiple (1996), p. 39.

[22]Seiple (1996), p. 150.

[23]In recent years, NGOs and the U.S. government have taken steps to improve coordination. InterAction—the American Council for Voluntary International Aid—was founded to improve coordination and professionalism among its members. With assistance from OFDA, InterAction is composed of over 150 U.S.-funded NGOs. It holds regular meetings and provides a place for the military and other government organizations to communicate with NGOs. Similar umbrella organizations exist for European NGOs, and several UN agencies also work with umbrella groups of NGOs that are common partners for them.

was announced.[24] Similarly, some NGOs assumed that the United States has superb intelligence on any crisis. U.S. officials' claims that they did not know where IDPs were or understand the local political situation were met with skepticism.

As a result of this limited familiarity, the military may not know who key relief partners and other important actors are in the early days of a crisis. As the USAFE after-action review of Support Hope noted, military personnel and the relief community "met on the dance floor."[25] Possible information sources are not sufficiently exploited both before and during a crisis. Before the intervention in Somalia, in-country NGOs were not asked to provide information. Similarly, U.S. personnel did not interview UN and NGO personnel before intervening in Rwanda. This failure to exploit available resources in Rwanda persisted during the intervention: The one intelligence representative in Kigali was also tasked with a host of other duties, including chaperoning visiting officials.[26]

LIMITED COORDINATION WITHIN NGOs

NGOs often do not coordinate well within their own organizations, leading to disjunctures during relief operations. The concerns of NGO field officers may differ considerably from those of their home agencies. Not surprisingly, field officers focus on day-to-day operations. At the national level, however, NGOs are concerned with pleasing their donors and maintaining a positive image for the overall organization.[27] Moreover, as noted above, the lack of an operational-level office for NGOs hinders coordination.

---

[24]Kennedy (1997), p. 105.

[25]United States European Command, *Operation Support Hope*, p. 3.

[26]Seiple (1996), p. 111.

[27]Dworken (1996), p. 16. The NGO operating environment also helps explain common differences between NGO headquarters staff and the field staff. Members of the field staff, prepared to face prolonged discomfort and personal risk, may be a different personality type than the home office staff, and they are likely to develop a different level of material and emotional involvement with the population they are helping. As with other undertakings and organizations, the view from headquarters is not necessarily the same as the view in the field.

Differences between NGO headquarters and field workers can decrease the benefits of previous NGO-military familiarization. Because of regular rotations and the large number of poorly trained, uninitiated personnel who travel to the field, agreements worked out with the main organization may not be carried out in the field. Aid organization officials who participate in exercises tend to be headquarters officials who seek to build long-term relationships rather than field workers.[28] Of all the NGO and UN staff, roughly 60 percent go into the field without any briefing. Often, this staff is recruited hastily, with little training or understanding of the NGOs' overall mission, let alone procedures worked out in advance to improve military cooperation.[29] Individual personnel come to rely heavily on their own instincts, and their own prejudices, in making decisions. For similar reasons, NGO officials in the field often lack the familiarity with the military that may have been painstakingly developed by NGO headquarters officials during exercises and by liaison staff in advance of a crisis.

UNCERTAINTY ABOUT THE MILITARY'S COMMITMENT

NGOs may be reluctant to invest in better coordination with the military unless they can foresee benefit. Most NGOs are small organizations with limited resources. Several interlocutors said that in the early 1990s they believed the U.S. military would often participate in relief operations during crises. They felt disillusioned when the United States decided not to participate or participated sparingly as during the Rwanda crisis. They hesitate to invest in exercises and planning, knowing that the U.S. government may not send its military to help after all.

Uncertainty leads NGOs to believe that any identity of interest between themselves and the U.S. military is likely to be situational and transitory.[30] In the next big crisis, whatever it may be, the United

---

[28]Dworken (1996), p. 31.

[29]Forman and Parhad (1997).

[30]Many NGOs also practice situational ethics, accepting military contributions while remaining hesitant to associate more closely with the military on general principles. There are situations—and they are becoming the rule rather than the exception— where the benefits of ties to the military are so essential that they will overcome any

States may not become involved. The NGOs, however, will most probably be there. As they see it, compromising their ability to function as neutral actors in a subsequent crisis is too high a price for better operations under a U.S. umbrella in a crisis.

IMPROVING PROSPECTS FOR COOPERATION

Several of the above problems have declined in severity in the last decade. Hostile stereotypes are falling, although they still interfere with cooperation. In the past, many military officers viewed NGO employees as young, antimilitary, self-righteous, incompetent, and unappreciative of security needs.[31] Their good intentions could produce disastrous results. As Jonathan Dworken notes, "Officers simply did not see women in their late-twenties with Birkenstock sandals and 'Save the Whales' T-shirts as experts worthy of consultation."[32] Our interviews suggest, however, a sea change in attitudes on both sides. Almost all NGOs and military officials noted their respect for the other and the need for consultation and cooperation. Almost all military officers who had worked with NGOs in crises noted their bravery and dedication.

Repeated interaction during crises and a decline in ideological tension after the end of the Cold War have helped reduce NGO suspicion of the military. NGO officials recognize that the military can respond to a crisis quickly and that, when U.S. forces arrive, they are ready to help the immediate relief effort. In addition, NGO members recognize that the military has made, and is making, a good-faith effort to improve its knowledge of NGOs and humanitarian relief problems in general. Several interlocutors noted that NGO officials

---

ideological qualms on the part of any NGO. The NGOs see no inherent contradiction in their position; other institutions often do. MSF refused the DoD offer to participate in the airlift for Hurricane Mitch relief but wanted the United States to provide aerial reconnaissance. To the military, this can look hypocritical: If you do not want to "corrupt yourself" through proximity to the military, you at least should be consistent. To MSF, their position is that they will accept help from the military only in an exceptional circumstance, an emergency. They had alternatives to the airlift, so they did not accept it. But when their helicopter went missing in Honduras, with medical personnel and a patient on board, the chance to save them overrode their scruples about requesting help from the military.

[31]Kennedy (1997), p. 109.

[32]Dworken (1995), pp. 19–20.

have far more respect for the military than they did just ten years ago—a sentiment corroborated by other interviews we conducted.

Growing concerns about security also are leading NGOs to shed some of their concerns about closer ties to the military. Almost all interlocutors noted that their organizations were far more focused on security than in the past and that they saw the military as a potential ally. Many NGOs report a lessening of respect for neutral parties present in a conflict, a breakdown of spoken and unspoken rules safeguarding helpers. MSF has had a number of doctors assassinated and seen its personnel and property targeted in Sudan, Afghanistan, Somalia, and Sierra Leone. It cites Iraq, the former Yugoslavia, Liberia, Chechnya, Rwanda, and Congo as areas where volunteers work under serious threat. Its activity report notes: "worldwide conflicts in which the impartial provision of humanitarian aid is less and less respected are becoming more common."[33] Concerns about evacuation in a crisis also are prompting many to seek better relations with the military.

These improving prospects for cooperation augur well for future NGO-military relations. If the military and NGOs are willing to implement procedural changes and devote resources to enhanced cooperation, overall performance in relief operations will improve. Several changes that would improve cooperation are presented in the final part of this report.

---

[33]Brauman (1993).

PART THREE. WORKING WITH EUROPEAN ALLIES

When the United States military provides relief during a complex emergency, it seldom acts alone: U.S. allies in Europe have a long tradition of humanitarian intervention. Several European militaries see humanitarian assistance as a future mission, and many have developed capabilities that can augment those of both the U.S. military and relief agencies. If anything, humanitarian intervention is becoming more important for most European militaries. NATO itself is playing an increasing role in these operations, partly because of the Kosovo crisis and partly because of a realignment of forces after the Cold War. The next chapter examines the European and NATO contribution, noting their capabilities and style of operations.

Chapter Ten

HUMANITARIAN INTERVENTION AS A COALITION ACTIVITY

European militaries are important partners for the United States in complex emergencies. European allies are, in general, more enthusiastic than Washington about humanitarian intervention and relief across borders, with or without the concurrence of host governments. The notion of the "right to intervene" has been a longstanding feature of French policy, and the Kosovo experience has strongly reinforced this tendency.[1] Britain under the Blair government has also espoused an activist approach to humanitarian assistance and intervention. Indeed, regardless of Europe's particular contribution to a complex contingency operation, its symbolic presence may be vital.[2] This connection has persisted in recent opera-

---

[1]Paris has argued that this humanitarian "droit d'ingerence" should be enshrined in UN practice and international law. The French government has been strongly supported in this approach by French-based NGOs such as Médecins Sans Frontières (MSF).

[2]The rationale for U.S. participation in humanitarian relief operations, including the provision of airlift, is often based on the perceived importance for *allied* interests. U.S. involvement, with Belgium, in the Congo crisis of 1960–1963 included the provision of U.S. airlift (for humanitarian as well as military logistics). Similarly, in 1964, the United States used C-130s deployed from Europe to support a complex humanitarian intervention by Belgian troops at Stanleyville. In the better-known Shaba I (1977) and II (1978) operations in Zaire, U.S. airlift was used to support French, Belgian, and Moroccan forces, alongside a regional peacekeeping contingent from Togo, Senegal, Gabon, and the Ivory Coast. Noncombatant evacuations and humanitarian relief were part of these operations. In each case, the rationale for U.S. involvement turned critically on the meaning of these contingencies for allied interests, especially Belgian and French stakes in Africa (alongside the implications for Cold War competition in the region). Mets (1986), pp. 121–136.

tions in sub-Saharan Africa and has been most evident in relation to humanitarian crises in the Balkans, where refugee pressures and the threat of violent spillovers touch directly on European interests.

The burdensharing dimension in relief operations is clearly seen in the overall security relationship with allies. It is especially true with European allies, where, in contrast to other types of power projection missions, the European contribution to humanitarian or complex operations can be substantial. In Africa, as in the Balkans, European contributions (measured in personnel and funding) to sustained relief operations often outweigh those of the United States.[3] This is an area in which allies "pull their weight."

EUROPEAN ACTIVITY AND OUTLOOK ON HUMANITARIAN MISSIONS

European allies have a tradition of humanitarian intervention in unstable regions, including Africa and the Balkans. The French experience is especially extensive (French perspectives are discussed in Appendix C). Humanitarian missions, as stand-alone operations and as part of complex contingencies, are becoming a more important part of European defense planning on a national and European Union (EU) level.

Apart from national contributions, the EU has come to play an active role in humanitarian assistance worldwide through the European Community Humanitarian Office (ECHO). ECHO has partnership agreements with over 170 NGOs and has particularly active working relationships with the Red Cross, MSF, Oxfam, Action Nord-Sud, CARE UK, GOAL, and Caritas. ECHO also administers a number of training programs in humanitarian and development assistance, principally through the EU's Erasmus program. ECHO's operations are worldwide, with the bulk of activity through 1996 (some 40 percent) focused on Africa, the Caribbean, and Pacific countries. In 1997, assistance to the Balkans was the largest item in the ECHO budget, a situation that is likely to continue for the immediate future. ECHO has also organized a number of country-specific task forces addressing crises in Haiti, Angola, Liberia, and elsewhere. Overall,

---

[3]Meier (1999), p. C1.

the EU is now the largest donor of humanitarian assistance world-wide.

ECHO has had representatives in all of the major complex emergencies of recent years; it also has its own offices in countries where the EU has representation. In the perception of close observers, the EU, through ECHO, has been moving to establish itself as an alternative to UN- and U.S.-funded activities, especially in Africa, often working through European-based NGOs. Relief efforts in Rwanda, Burundi, Guinea, Ivory Coast, and Mali have been heavily European in funding as well as logistics.

Beyond the role of ECHO, the EU is emerging as a more concerted actor in relief operations as a dimension of foreign and security policy. Peacekeeping, peace support, and humanitarian missions figure prominently in defense restructuring across Europe. An enhanced capability for expeditionary operations—part of this process—is highly relevant to humanitarian response outside Europe and on Europe's periphery. Indeed, the early warning and crisis-management mechanisms being developed within the EU (in support of the Common Foreign and Security Policy—CFSP) are likely to focus on new areas of potential humanitarian emergency beyond Africa and the Balkans.

The Kosovo experience has had a transforming effect on discussions of future EU foreign and security policy. The EU already had a significant commitment to humanitarian activities, through ECHO and through an earlier decision to allow the Western European Union (WEU) to coordinate European humanitarian interventions (these WEU functions are now likely to be fully absorbed by the EU).[4] Kosovo greatly accelerated EU involvement and has led to closer consultation and coordination between the EU and UNHCR. The nature of emerging European defense capabilities and the less-controversial nature of humanitarian military deployments suggest that Europe's new CFSP will have a strong humanitarian dimension. At the same time, European defense initiatives (ESDI) are likely to give European allies an even greater capability for expeditionary operations in humanitarian crises. Although some have suggested

---

[4]Kuhne, Lenzi, and Vasconcelos (1995).

that certain European countries are "free riding" on the military side of the Balkan security equation by opting for "soft burdens" such as aiding refugees, this ignores the reality that much of Europe's humanitarian role in the region is being carried out with military assets and personnel.[5] Finally, these trends are also encouraging some of the most active players in Africa, especially France, to channel more of this activity through European structures and organizations.

THE MILITARY DIMENSION

European allies, especially France and Belgium, have been willing to place military assets at the service of humanitarian relief operations in Africa, the Balkans, and farther afield. It may involve the outright donation of airlift, without expectation of reimbursement by NGOs or UN organizations—something the United States has traditionally been reluctant to do (the cost of "shared" U.S. lift often comes as a surprise to European militaries and NGOs). Examples include the use of French and Belgian military aircraft in support of operations in south and central Somalia and in Sudan. U.S. observers have been especially impressed by the activity and professionalism of the Belgian military in Africa, and there has been routine coordination among U.S., Belgian, and French personnel in humanitarian training and relief flights. Peacekeeping and humanitarian relief deployments to East Timor also have a substantial European component. Portugal, with its colonial ties to the province, deployed 1000 troops, along with two frigates, C-130 transports, and four helicopters. France, Canada, Italy, Sweden, and Finland also participated, with a heavy emphasis on humanitarian assistance.[6] European allies are generally more willing than the United States to transport NGO personnel on military aircraft.

In Africa, European airlift activity can be discussed in terms of three rough categories—most active, moderately active, and limited or specialized activity. In the first rank, France and, to a lesser extent, Belgium are routine participants in relief operations with a military airlift component. A second-tier group of moderately active states

---

[5]Germany has been cited as an example of this humanitarian-based free riding. "Guns or Refugees—an Unequal Alliance?" (1999), p. 50.

[6]Associated Press (1999).

includes Italy and the UK. A larger, third group consists of Portugal, Germany, the Netherlands, Canada (included as a "European" ally for the purposes of this discussion), and Spain. Overall, European militaries are providing some 10 percent of current African humanitarian airlift. The vast bulk of airlift requirements for routine and emergency assistance in Africa is provided by private cargo charters. Even in this case, European companies are at the forefront, and there are large numbers of Russian and Ukrainian aircraft.[7]

European allies do not possess the heavy lift and extensive command and control assets of their U.S. counterparts and have a more limited ability to operate on a remote basis. But these shortcomings, evident in many transatlantic defense comparisons, do not necessarily weigh heavily in relief operations. Heavy lift may not be necessary in relief activity, particularly after the early days of a crisis, and airlift in general is an expensive option if sealift and overland shipment are possible. Where heavy airlift is required, commercial vendors are the routine alternative. Moreover, these vendors can and do operate without the operational restrictions imposed by safety and security on military airlift.[8] In many cases, tactical airlift is most useful for supporting relief operations in austere or insecure environments, and European allies possess significant tactical lift assets.

REGIONAL BASES AND EXPERIENCE

Former colonial connections and continuing defense relationships do not always make for easy political relationships between Europe and local actors in Africa. But in a narrower operational sense, this familiarity confers advantages. France and Belgium have linguistic and cultural ties across much of West and Central Africa. Portugal

---

[7]African companies are also active. Transafrique, SAFAIR, and SAT (Southern Air Transport) are among the largest operators. Transafrique reportedly has lost aircraft over Angola, probably from ground fire. Of the roughly 25 incidents of shoulder-fired surface-to-air missile (SAM) attacks on civilian aircraft, almost all have occurred in sub-Saharan Africa. During the Sarajevo airlift, an Italian military transport was reportedly hit by a ground-to-air missile.

[8]For example, in Operation Provide Hope, U.S. airlift from Mombassa to southern Somalia allowed for higher delivery volumes than might otherwise have been possible, but the USAF was constrained in its operational mandate. Under similar conditions, private air cargo vendors might fly three or four rotations per day.

has similar ties to Angola and Mozambique. Italian ties are strong in Somalia. In many cases, military-to-military ties are well established. In the French case, the maintenance of bases with forward-deployed mobility forces around Africa offers advantages for military intervention and humanitarian relief. Bases on the African periphery (Dakar, Djibouti, Reunion, the UK base at Ascension Island, and Spanish facilities in the Canary Islands) are useful for lift into Africa from Europe.

Local experience is another European capability element. For allies with defense ties to Africa (above all France, but also Spain, Belgium, Portugal, Italy—and Israel), it is not unusual for some military personnel to have spent their entire careers in Africa or involved in African operations. As a result, these military establishments have useful niche capabilities in, for example, tropical medicine, civil engineering, and intelligence. Even where there has not been an historic connection, allied governments will sometimes find it useful to deploy specialized military assets for humanitarian and political reasons, as with the prominent 1994 deployment of an Israeli Defense Forces field hospital to Goma during the crisis in Rwanda. In this case, the Israeli deployment, including airlift from Israel and security, was supported entirely with Israeli assets. [9]

RELATIONS WITH NGOs AND NGO CONCERNS

The problem of civil-military relations in relief operations is a concern for European military establishments, just as it is a concern for U.S. military planners. European NGOs are, by and large, wary of cooperation with the military, although the trend toward ever-closer interaction is a phenomenon affecting NGOs and militaries worldwide.

However, observers of operations in Africa and the Balkans where both U.S. and European militaries have been engaged note some differences in approach to NGOs. In some instances, European militaries appear to have a better liaison relationship, especially with European NGOs. Indeed, each major country has its own favored

---

[9]This deployment and the decisionmaking process behind it are discussed in Wiener (1997).

"core-team" NGOs, as does the United States. There is often a closer, more collegial "corporatist" approach, and less friction over force protection issues (the relationship between the French military and NGOs may be an exception to this). Some factors that may contribute to an easier civil-military relationship include greater local knowledge, language ties, and the fact that many European militaries view relief operations as a core mission. Local diplomatic staff often play a key liaison role (Belgian and French embassy staff have been described as "omnipresent" in recent African operations). European military establishments routinely assign liaison officers to key NGOs and UN agencies.

European NGOs offer perspectives on their mission and current challenges that roughly mirror the views of NGOs based elsewhere. There is a strong perception that when the U.S. military and NGOs are involved in relief operations, they bring enormous means but disdain close contact with the population. One French NGO official (in an observation repeated by several French military officers) recalled the impression of an "invasion" created by U.S. relief operations in Africa, with personnel in battle dress. Spanish, French, and other European militaries have, by contrast, tried to get "closer to the people." European NGO personnel also tend to regard UN agency staff as equally distant from local conditions and highly bureaucratic.[10]

At a philosophical level, it is noted that NGOs are uncomfortable with the military, but in the field there is often effective cooperation. Interaction with the military is set to grow because European NGOs are increasingly concerned about personnel security in relief operations. The French Red Cross alone gives security training to some 60 people per year, and such training is now common in other European Red Cross organizations.[11] Despite the growing role of private security organizations in relief operations, these arrangements are not popular with European NGOs.

---

[10]European NGOs that are especially active in African relief include Médecins Sans Frontières, ACF (Action Against Hunger), national Red Cross organizations (particularly in France, Spain, Sweden, and the UK), Médecins du Monde, and Caritas France.

[11]The growing threat to UN and NGO relief workers is highlighted in Miller (1999).

The NGO community in Europe is now trying to look ahead to emerging areas of need such as the Caucasus and Central Asia (especially Afghanistan). The demands for humanitarian assistance in these areas are potentially large, and the environment for intervention is among the most insecure, with threats of kidnapping and mines looming large. Airlift is another concern. The extensive reliance of European NGOs on the air charter market is troubling because of its inherent cost (and the rates rise as NGOs compete for space in emergencies) and poor safety and security offered by air charter operators, especially those based in eastern Europe and the former Soviet Union.[12] As a result, European NGOs are exploring the possibility of maintaining dispersed or prepositioned stocks close to likely crisis areas in Africa and the Pacific (e.g., at Madagascar, Mauritius, the Seychelles, Mozambique, and New Caledonia). In some cases, these decentralized facilities might be colocated with existing French or other military bases. All of these issues, from risk assessment to logistics, suggest to European observers the need for better geopolitical analyses by NGOs and within relevant EU organizations.

TRANSATLANTIC SYNERGIES IN HUMANITARIAN CONTINGENCIES

In those instances where U.S. and European militaries have worked together in evacuation and humanitarian assistance operations, European officers have generally been impressed with the capability (especially airlift and associated ground-handling operations) and professionalism of U.S. forces. In large-scale humanitarian relief operations, as in central Africa, Somalia, and the Balkans, there is an important potential synergy between European and U.S. assets and capabilities. European allies have extensive local ties, useful tactical lift, and training for humanitarian interventions. The United States can offer heavier lift to the African theater, big-picture intelligence, and command, control, communications (C3) assets, without which large-scale operations may be delayed or ineffective. Not least,

---

[12]As an example, in one recent case the French Red Cross spent $100,000 to fly 36 tons of materiel to Brazzaville.

European allies, through the EU, are also in a position to provide substantial funding for humanitarian relief.

WHAT ROLE FOR NATO?

Even before the Kosovo crisis, NATO was moving steadily toward a more active role in humanitarian-related missions. Kosovo, the articulation of a new Strategic Concept, and operational developments in the Balkans have accelerated this trend and created new patterns of relations among NATO, UN organizations, and NGOs. The net result is that NATO, once a marginal actor in relation to relief operations, is now an important vehicle for civilian and military planning in this sphere.

Although NATO's role is still in transition as a result of the Kosovo crisis and ongoing activities in the Balkans, the role of Kosovo in changing NATO's perspective on humanitarian missions, especially airlift, has been pronounced. NATO's structures for civil emergency planning, the control of air movements, and standardization in logistics, all of which are useful in relief operations, were underdeveloped. The civil emergency planning focus in NATO has traditionally been oriented toward managing refugee crises behind a conventional front in western Europe or reconstitution after nuclear attack. Although SHAPE (Supreme Headquarters Allied Powers Europe) had mechanisms for managing airlift flows, airlift itself has always been a national responsibility within the Alliance (i.e., members brought their own airlift assets to NATO operations). There is now discussion about NATO acquiring new heavy lift assets through the infrastructure fund, but its outlook is unclear.[13]

The NATO approach toward humanitarian missions draws on several elements. First, NATO has a long-standing interest in and structure for civil emergency planning. In the post–Cold War environment, NATO's Civil Emergency Planning activities have become a useful vehicle for engaging Partnership for Peace (PfP) and Euro-Atlantic Partnership Council (EAPC) countries, Russia, and Mediterranean

---

[13]Individual NATO allies, including Germany and the UK, also have plans to acquire strategic lift, either by charter arrangements (e.g., with Ukraine) or by purchasing airlifters (C-17s or the "Future Large Aircraft").

dialogue states in noncontroversial forms of cooperation. With the formation of the Euro-Atlantic Disaster Response Coordination Center (EADRCC) within the Alliance's Civil Emergency Planning Directorate in June 1997, NATO has been able to play a more significant role in responding to natural disasters and humanitarian crises of various kinds. The EADRCC has been a key actor in refugee relief in Kosovo and in coordinating allied contributions to earthquake relief in Turkey and Greece. UN agencies are now a leading consumer of EADRCC efforts, which focus on improving and coordinating the emergency response capabilities of the 44 members of the EAPC.[14]

A second stream of NATO humanitarian engagement has been the growing focus on peacekeeping and peace support activities, both as Alliance missions and as focal points for training and cooperation with partners. In Bosnia, and more recently in Kosovo, these missions have unfolded in a complex setting in which humanitarian relief has been a key task. The new NATO Strategic Concept, presented at the 1999 Washington Summit, refers prominently to crisis response missions and capabilities, including disaster relief and humanitarian assistance.[15] The Mediterranean Cooperation Group, the Alliance body responsible for activities within the NATO dialogue with nonmember states in North Africa and the Middle East, has also identified civil emergency planning, including humanitarian relief, as a subject for future initiatives.[16]

The EAPC's Ad Hoc Group on Cooperation in Peacekeeping has been a focal point for Alliance thinking, especially on the humanitarian and civil-military aspects of peacekeeping and peace support operations. NATO, in cooperation with national governments, has organized meetings with international organizations and NGOs to discuss information exchange and predeployment preparation in relief operations, most recently in Geneva in February 1999.[17] Areas of

---

[14]For a summary of NATO civil-emergency planning activities and issues (pre-Kosovo), see Palmeri (1998); Palmeri (1996).

[15]Specific references to humanitarian missions appear in both the Washington Declaration, NAC-S(99)63, pp. 3, 9, and 13; and the new Strategic Concept, NAC-S(99)65, pp. 3, 12, and 16 (see North Atlantic Council in the bibliography).

[16]See Lesser et al. (1999).

[17]See *Cooperation in Peacekeeping: Workshop on Humanitarian Aspects of Peacekeeping* (1999), and Klaiber (1999). Participants included representatives of UNHCR,

emphasis in recent discussions include the need to develop a single focal point for planning, a civil-military exercise program, and standardization in policy, doctrine, and rules of engagement. In addition, there is interest in how the restructuring of defense forces in Europe can support future relief operations. The theme of civil-military cooperation in humanitarian airlift has also been taken up by the NATO Air Defence Committee (NADC).[18]

The Ad Hoc Group is compiling a compendium of views on the humanitarian aspects of peacekeeping. This open-ended document will not have any formal standing in determining NATO doctrine but will be influential in shaping NATO practice. Its content is now likely to be strongly influenced by the perceived lessons of the Kosovo experience. The document, pre-Kosovo, summarizes well-known observations and concerns on complex relief operations and the problems of civil-military cooperation. It identifies the key international organizations with whom coordination will be essential (it acknowledges but does not attempt to catalogue the range of relevant NGOs). The compendium draws a useful distinction between civil-military cooperation at the political/strategic, theater, and tactical/field levels, reaffirms the lead role of humanitarian agencies, and strongly endorses the inclusion of NGOs in peace support exercises. It stresses the need for civilian humanitarian actors to have direct contact with operational staffs (rather than with public relations officers) and accepts the notion of liaison officer exchanges in military and NGO field headquarters. With the Bosnian experience in mind, UNHCR and other humanitarian personnel have been participants in training courses and seminars at the NATO school in Oberammergau, as well as activities organized by the North Atlantic Cooperation Council, now EAPC.[19]

As NATO implements its Combined Joint Task Force (CJTF) concept, it has been suggested that NATO consider establishing "humanitar-

---

ECHO, ICRC, UNICEF, OSCE, the International Committee of Voluntary Agencies (ICVA), Supreme Allied Commander, Europe (SACEUR), and Supreme Allied Commander, Atlantic (SACLANT).

[18]An NADC seminar, "Coordination and Cooperation Between Civil Administrations and Air Defence During Humanitarian and Peace Support Operations," was held at the French Air Force Academy, October 6–9, 1998.

[19]Mendiluce (1994), p. 26; Lightburn (1996).

ian CJTFs" that would allow for the participation of NGOs and IOs alongside national contributions, much as the current CJTF concept provides for the participation of nonmember states in NATO-led military operations. Given the desirability of civilian control in relief operations, and the desire of NGOs for access to NATO information while safeguarding their independence, the usual practice has been to establish parallel and coordinated—rather than unified or "joint"—planning and coordination mechanisms. It is therefore unclear whether a humanitarian CJTF could be made to work in practice when NGOs are involved in great numbers. Relief efforts in Kosovo will be an important test case in this regard—the KFOR (Kosovo Force) Combined Joint Civil-Military Task Force is, in essence, a humanitarian CJTF.

In the debates over the Strategic Concept and future NATO missions, there have been clear differences in perspective among allies. Canada, the Netherlands, Belgium, and others whose defense doctrines already emphasize humanitarian missions have favored the emergence of relief operations as a core task for NATO. But others, principally the UK and France, have been ambivalent. British policymakers, while not averse to humanitarian intervention (as the Kosovo experience makes clear), do not wish to see NATO resources drawn away to address humanitarian missions. France, in contrast, strongly favors humanitarian missions as a matter of national strategy, but NATO is not its preferred vehicle for action, for long-standing political reasons. The new members, and the PfP states, especially neutrals, have been most enthusiastic.

Logistics, including airlift, is a third subject of NATO relevance to relief operations. Standardization in logistics has been a long-standing NATO interest, and considerable attention has been devoted to problems of efficiency and harmonization in this area.[20] Although the provision of airlift for military or humanitarian purposes is a national responsibility in NATO, SHAPE has procedures for pooling airlift and managing traffic flows to promote efficiency. The IFOR and SFOR operations involved a large pool of this sort. SHAPE planners stress that the key issue in these complex operations has been

---

[20]The detailed NATO guidelines on logistics practice include humanitarian missions as part of peace support operations. See Senior NATO Logisticians' Conference Secretariat (1997).

the movement of people rather than materiel. There has been some sharing of space with NGOs on a space-available basis; NGOs have, in turn, offered space to U.S. personnel on their chartered aircraft. In Balkan relief operations prior to Kosovo, NGOs have typically turned first to national governments rather than to NATO when seeking airlift. When there are approved requests to NATO, SHAPE then polls nations on available lift.[21] Several NATO allies (e.g., the Netherlands, Belgium, and Canada) have traditionally been more willing than the United States to donate tactical lift. When requests come through the UN, the UN will generally purchase lift space. NATO's Civil Emergency Planning directorate is somewhat unusual in that it can fill lift requirements for humanitarian response or disaster relief directly, through commercial charters, without reference to SHAPE.

Another resource is provided by the standing Civil Military Cooperation Cell (CIMIC) at SHAPE. The cell is responsible for assessing the effect of military operations on the civilian population and environment, and vice versa. Even before Kosovo, CIMIC was the lead vehicle for NATO's military interaction with international organizations and NGOs. In this regard, the operational-level interaction through CIMIC was generally more active than equivalent contacts at the political level through NATO Headquarters. In response to the Kosovo crisis, CIMIC established a Refugee Support Coordination Center (RSCC) to coordinate military support for civil humanitarian relief operations in the Balkans. Recent NATO debate on civil-military relations has pointed to the need to extend CIMIC-like arrangements that have worked well at the operational level to embrace higher-level interactions at the political level, bringing together NATO decisionmakers with other key international organizations (and perhaps NGOs) active in peacekeeping and relief operations.[22] The Kosovo experience arguably brought this about through crisis-driven necessity.[23]

---

[21]Through the Allied Movement Coordination Center (AMCC).

[22]Zandee (1999), pp. 12–13.

[23]Civil-military coordination in Kosovo will be addressed in greater detail in future RAND work.

SOME OBSERVATIONS ON KOSOVO

The 1999 Kosovo crisis has greatly advanced the humanitarian dimension of Alliance strategy and planning. Although responsibility for humanitarian activities was vested with UNHCR, it quickly became clear that NATO would have to play a more direct role. UNHCR, despite its nominal claim to be a "lead agency," had neither authority nor resources in Kosovo. NATO's relief operations took place on many fronts, including (1) managing the airlift of relief supplies, (2) temporarily transferring some refugees to NATO countries, (3) off-loading and providing immediate storage of relief supplies, (4) building refugee facilities, and (5) attempting to provide information on the number and location of internally displaced persons.[24]

The NATO role in coordinating humanitarian relief in the Kosovo crisis was greeted with some ambivalence in the NGO and IO communities. There was disappointment at the inability of UN agencies to manage the relief effort and skepticism about NATO. One NGO official described the NATO involvement as "partly scraping the bottom of the barrel" and NATO "trying to establish a new role for themselves."[25] Reactions of this sort may have flowed, in part, from perceptions early in the crisis that NATO would not mount a serious campaign to restore the autonomy of Kosovo. That said, it appears that the relationship between NATO officials and NGO and IO representatives was effective and generally positive, despite the challenging circumstances and the disproportionate capabilities in the field.[26] It is also important to recall that the NATO-UNHCR relationship did not begin with Kosovo but began in earnest with operations in Bosnia, with UNPROFOR (the UN Protection Force), and the Sarajevo air bridge in which NATO allies played a key role.[27]

Although the lessons of the experience from all sides need to be assessed more closely, it appears that NATO's foray into humanitarian

---

[24]Balanzino (1999), p.10.

[25]Quoted in Kaminski and Copetas (1999).

[26]The UN's 1999 Interagency Needs Assessment Mission for the Federal Republic of Yugoslavia apparently did not include a NATO representative among its numerous participants.

[27]Mendiluce (1994).

relief was conducted with unprecedented attention to civil-military relations and liaison arrangements. In KFOR, the civil-military relations effort is being carried out primarily through a Combined Joint Civil-Military Task Force (CJCMTF) working directly for the KFOR commander. The CJCMTF has the responsibility for liaison with NGOs and IOs, with the eventual aim of returning the bulk of humanitarian relief efforts to civilian control.[28]

The unprecedented NATO involvement in relief operations in the Balkans, and later in the Kosovo crisis, has a number of implications. First, it is clear that humanitarian support has been firmly established as a future mission for the Alliance. Second, it is now more likely that coalition approaches to humanitarian relief within the Euro-Atlantic area will be coordinated through NATO rather through ad hoc multilateral arrangements, although there will still be a strong tendency for military-NGO interaction along national lines. This tendency was evident in Kosovo, as well as in IFOR/SFOR, and is reinforced by national contingent deployments operating within national sectors. Third, NATO has emerged as an important stakeholder in humanitarian planning and consultations—alongside UN agencies, the ICRC, and key NGOs—and as a new player on issues such as "codes of conduct" for NGOs. Closer NATO/EU/UNHCR coordination has been a feature of the Kosovo experience. Fourth, within NATO, the Kosovo experience has undoubtedly strengthened the role of civil emergency planning and, in particular, the EADRCC.

Thus, the Kosovo experience will likely reinforce underlying trends for a more active NATO role. The growing focus on power projection and expeditionary operations, even for traditional NATO missions, will mean more (and more widely shared) inherent capability for remote relief operations. More flexible command arrangements, such as the CJTF concept, will have particular relevance for humanitarian interventions, as will the focus on peacekeeping and peace support operations. Finally, new and prospective Alliance members and partners tend to be among the most interested players in relief operations; they will bring important niche capabilities, and future training exercises with their militaries will likely emphasize peacekeeping and humanitarian assistance.

---

[28]Balanzino (1999), p. 13.

PART FOUR. IMPLICATIONS AND RECOMMENDATIONS

The U.S. military, and the U.S. government more broadly, cannot solve all the problems inherent in complex emergencies. They can, however, minimize the impact of many of these problems while leveraging NGO, IO, and allied capabilities more effectively. Although our recommendations emphasize what the USAF and the unified commands should do, many of them require action across all the military services, the unified commands, and the Department of Defense, as well as by key civilian agencies, such as the Department of State.

Solving the most difficult problems would require a major resource commitment by the U.S. government. Such a commitment does not appear to be likely, so Chapter Eleven focuses on more limited steps, including organizational and operating procedural changes that can increase U.S. effectiveness in humanitarian assistance.

A STRATEGY TO IMPROVE COORDINATION

This chapter outlines a strategy to improve coordination with relief agencies during humanitarian crises, including suggestions for dividing up responsibility for implementing the strategy. It identifies the advantages of such a strategy as well as potential difficulties. By implementing this strategy, the military will be better able to take advantage of relief agency capabilities and minimize problems.

More effective provision of relief requires overcoming or minimizing many of the problems that currently affect cooperation between the military and relief agencies and capitalizing on U.S.-allied synergies. A strategy to improve coordination would have the following objectives:

- Ensuring familiarity with relevant relief organizations.

- Improving information sharing both before and during crises.

- Fostering better long-term planning and coordination by closely engaging select relief organizations.

- Improving coordination of the relief flow during humanitarian crises.

- Encouraging developments among U.S. European allies to improve their humanitarian relief capabilities.[1]

---

[1]These objectives overlap and reinforce one another. Increasing familiarity and engaging key NGOs will ease the coordination of the airflow during a crisis. Similarly, better information sharing will strengthen the overall engagement effort.

Meeting these objectives requires both institutional changes in the U.S. military at multiple levels and a change in procedures for carrying out relief operations.

At a minimum, the military should ensure that its key personnel are familiar with organizations relevant to relief operations. At the same time, it should help these same organizations become more familiar with the military's organization and capabilities. Relevant organizations would include several agencies in the UN family, the ICRC, and a broad spectrum of NGOs. Familiarization should promote mutual understanding and better cooperation across the military, UN agencies, and NGOs.

In addition, the military should closely engage select organizations that play key roles during humanitarian crises in order to improve long-term planning. Key organizations would include agencies in the UN family (e.g., OCHA, WFP, UNHCR), the ICRC, and selected NGOs, particularly the core-team NGOs identified earlier. Engagement would speed response and increase efficiency during all phases of a humanitarian crisis, especially during the initial phase when delay might cost lives.

Building on both these efforts, the military should initiate actions to improve coordination of the relief flow during humanitarian crises. The services should offer their impressive logistics capabilities to help manage the airlift and sealift of supplies, particularly in the early days of a crisis. To be fully effective, these actions should address both the narrower problem of managing the aid flow and the broader, more fundamental problem of establishing priorities.

The military should also recognize the important role that European allies can play in responding to complex emergencies. The United States should encourage European militaries to further develop their capabilities in this regard. Equally important, the U.S. military should improve its ability to leverage these capabilities and augment them with its own.

The particular requirement will determine which military element should act to achieve the objectives. The unified command is the most appropriate entity to carry out many of the most important actions recommended below. Almost all of the recommendations apply to the regional commands (i.e., USEUCOM, USCENTCOM,

USPACOM, and the USSOUTHCOM, but several key recommendations apply to functional commands, particularly the U.S. Transportation Command (USTRANSCOM). Several vital steps, however, require the armed services, the Joint Staff, or DoD to play a leading role. When appropriate, the remainder of this chapter links specific recommendations to appropriate elements within the military.[2]

The military cannot promote coordination alone. An effort to engage NGOs and improve the flow of aid requires not only cooperation among the organizations identified in this report but also among donor and host countries at high political levels. This is particularly true regarding steps to improve the capabilities of European allies. But the military can improve performance by identifying the problems, advocating workable solutions, and promoting solutions before crisis occurs.

ENSURE FAMILIARITY: RECOMMENDED ACTIONS

Greater familiarity would promote mutual understanding between the military and relief organizations and reduce lingering suspicions of the military within some NGOs. It also would help the military take advantage of expertise resident in the NGOs and smooth coordination during a crisis. The military should become familiar with all NGOs operating during a crisis, particularly those belonging to the "core" category. Although minor and advocacy NGOs may contribute relatively little to the overall operation, their personnel nevertheless could be taken hostage, threatened, need transport, or otherwise require military assistance. Moreover, an otherwise minor NGO may play a major role in a particular contingency.

The following initiatives, if taken by the unified commands (and, to a lesser degree, the armed services, and other actors in the defense community) would help ensure greater familiarity:

---

[2]Because of the role of USEUCOM in initiating this research, many of the recommendations for implementation at the unified command level use USEUCOM as the example. Except where otherwise noted, we believe these actions would be beneficial in other commands as well.

- Appoint a "humanitarian advisor"
- Systematically and routinely brief relief agencies on military capabilities
- Integrate civil affairs capabilities into noncrisis operations
- Sponsor conferences and seminars
- Sponsor partnership with the Center of Excellence.

A more detailed division of labor for implementing these steps is suggested in Table 11.1 (pp. 152–153).

Appoint a Humanitarian Advisor

To ensure better familiarity with relief agencies—perhaps the biggest step to improving overall coordination—an individual should be appointed by each unified command to work with NGOs and IOs. (If a unified command or the military in general seeks to emphasize this mission beyond current levels, the appointment of additional individuals should be considered.)

The military generally is unfamiliar with other actors during humanitarian crises. Many officers have some knowledge because of their participation in previous relief efforts, but there is little effort to maintain regular contact or ensure institutional awareness of relief agencies. Although military officers are broadly familiar with the role of ICRC in implementation of the Geneva Conventions, their normal duties do not require them to become familiar with the UN family, NGOs, or the ICRC in its relief capacity. With few exceptions, military officers are not trained to work with these organizations. Joint doctrine identifies important NGOs and sketches their capabilities, but only in a generic fashion.[3] USEUCOM and other unified commands need to know where NGOs are working within their AOR and be at least broadly acquainted with their programs and capabilities for quick response.

For example, no staff entity in USEUCOM currently has a responsibility to ensure that the command is familiar with those NGOs that

---

[3]Joint Chiefs of Staff (1996), Joint Pub 3-08, pp. D-3 and D-4.

are working within the AOR and those that would likely arrive during a crisis. Some staff do occasionally work with NGOs and IOs, but not as their primary responsibility.

Although U.S. embassies and country teams are often knowledgeable, unified commands cannot count on them to provide information during crisis. Many embassy officials interact with relief agencies from time to time, and some are highly knowledgeable about relief activities. Embassies' primary responsibility, however, is to conduct relations with the host government, leaving them less familiar with NGOs and local conditions outside the capital. Even U.S. embassies within the AOR may not be fully informed or appreciate the unified commands' need for information concerning NGOs. Some embassy country teams are fully informed of current NGO activities, but many are not. Particularly in Africa, embassy personnel are often responsible for multiple countries and are restricted in their access, making them unable to work closely with aid organizations. The USAID representative in an embassy is cognizant of NGO programs sponsored by the U.S. government but not necessarily about efforts sponsored by other governments. The defense attachés in Africa may have little direct contact with NGOs or UN agencies. Moreover, the defense attachés are often associated with host nation military and security personnel, which NGOs may see as part of the problem. Thus, though many country teams are valuable resources, they are not consistent in their knowledge of NGOs and other relief agencies.

To improve its ability to coordinate with NGOs, each unified command should designate a humanitarian affairs advisor—a "HUMAD"—as an individual responsible for crisis liaison with relevant agencies in the UN family and NGOs in the AOR.[4] This individual should be able to offer NGOs access to the command's resources; otherwise, NGOs may feel that the liaison is a one-way street. This same individual should have a working knowledge of relevant agencies in the United Nations. To assist this individual, the command should encourage country teams and defense attachés to track NGO activities and report on them.

---

[4]The Joint Commanders-in-Chief wargame on complex operations also recommended the creation of a HUMAD comparable in status to a CINC's political advisor.

The HUMAD should develop personal contact with NGOs. Because NGOs are comparatively nonbureaucratic, their personnel respond better to personal relationships than to institutional ties. All NGO officials interviewed stressed the importance of personal relations—"We want someone in our Rolodex to call," noted one aid official. They will provide information more willingly and in greater detail to an individual known to them than to a faceless organization. The HUMAD should also track unified command personnel with experience in complex emergencies and know which individuals have contacts with relief personnel.[5] During crises, the HUMAD should be the command's primary point of contact with NGOs. The HUMAD might deploy with early arriving forces to help set up a CMOC and otherwise ensure orderly coordination.

Brief Relief Agencies on Military Capabilities

The military can also improve familiarity by briefing relief agencies on the military's capabilities, limits, culture, and procedures. Regular briefings of NGO personnel conducted by the Joint Staff (for U.S.-based NGOs) and the appropriate command elements (for important NGOs active in the AORs) and briefings by relevant service components, such as Air Mobility Command (AMC), would help the relief community gain a more realistic picture of the military.

Most other actors in humanitarian relief operations, especially the NGOs, know even less about the military than the military does about them. With the exception of some retired military officers working for relief agencies, few NGO personnel have experience with the military other than occasional glimpses during major crises. They are bewildered by military organization: They are unacquainted with the unified command structure, know almost nothing about the workings of joint staffs, and do not understand military command and control. They have an unrealistic picture of military capabilities, tending in general to overestimate what the military can accomplish. For example, they think that the military can deploy in

---

[5]Other staff officers concerned with relief operations can expand their knowledge through training offered by NGOs and academic instruction. The USAF and other services can provide information on NGOs, civil affairs capabilities, and the UN system in service schools.

days or even hours when in fact weeks are required, or they may not understand the limitations of intelligence sources such as overhead imagery. Several NGO officials believed that the U.S. military was lying when it claimed that it could not determine the location of refugees or that bad weather interfered with intelligence collection.

NGOs should become familiar with the military, preferably before a crisis begins, when time is less critical. They should understand enough about military organization and military command and control to facilitate coordination. They need to know where to turn for specific purposes and to understand how the military handles requests for support. They need to know the functions of a CMOC and to appreciate its place in the command and control structure. They need a general appreciation of aerial port operations.

Coordination would also be easier if NGOs appreciated what the military can and cannot do. In past crises, however, even core-team NGOs tended to credit the military with unrealistic capabilities. They assumed that the military could secure their highly dispersed operations, that it could easily disarm combatants, or that it had reliable intelligence on refugee movements. NGOs do not have to become expert in military operations, but they do need more realistic expectations.

Several channels are available to inform other actors, especially NGOs, about the military. Unified commands such as US-TRANSCOM and service components such as AMC and U.S. Air Forces in Europe (USAFE) can prepare and distribute materials. Before an operation, DoD, the Joint Staff, or a unified command should brief NGO representatives. Days before the intervention in Haiti, for example, U.S. Atlantic Command (USACOM) briefed chief executive officers of NGOs concerning the operation. If a Center of Excellence (discussed below) were established, it could mount a continuous, ever-widening effort to familiarize NGOs with the military. As noted below, conferences, seminars, and exercises could also contribute to mutual familiarity.

In all such efforts, the military should strip away extraneous verbiage, including catchwords and self-advertisement. It should keep abbreviations and acronyms to an unavoidable minimum and explain them at every fresh use. It should tailor briefing materials to the mission, i.e., telling NGOs and other actors only what they need to cooperate smoothly with the military.

Integrate Civil Affairs and Other Specialists into Noncrisis Planning

The unified commands can draw on Civil Affairs and Special Forces personnel to ensure familiarity with NGOs.   During Operation Provide Comfort, for example, these personnel established a rapport and close working relationships with NGOs.[6] Special Forces personnel are trained to work with civilian agencies and have personal acquaintance with local conditions.   Civil Affairs personnel, particularly within the Army, usually have broad familiarity with NGOs and understand their roles in relief operations.[7]

For precrisis planning, these assets may be of limited utility. The Army has only one active-duty Civil Affairs battalion located at Fort Bragg. All other Civil Affairs assets are in the inactive components and may not be called into active duty in time to participate in crisis response. Because many are located in the reserve forces, they are frequently unavailable for precrisis planning or in the early days of a crisis. Perhaps most important, Civil Affairs and Special Forces personnel are often overextended, given the high demand for peacekeeping operations.

Several steps would allow the military to gain more benefits from Civil Affairs specialists. Expanding Civil Affairs and placing additional units on active status would enable the unified commands to draw on their expertise as needed before a crisis begins. Individuals from these units could then deploy with early arriving forces to ensure smooth coordination. If this status cannot be changed, the commands must more aggressively call upon Civil Affairs personnel in the planning stage, recognizing their potential contribution to these operations. Special Operations forces should be invited to planning meetings, exercises, and other activities that will involve cooperation with relief agencies.

---

[6]Seiple (1996), pp. 22–23.

[7]Natsios (1995), p. 79.

Sponsor Conferences and Seminars

Conferences and seminars can be used to familiarize military participants with various agencies and techniques to improve their cooperation or coordination. To be effective, they should be organized
around topical themes of mutual interest. The NGOs should represent a spectrum that ensures participants will hear new information
and encounter fresh perspectives. If possible, they should include
representatives from NGOs that have shown little inclination to seek
contact with the military, such as MSF or other core-individual
NGOs. They should include key agencies of the United Nations and
the ICRC. It will usually be easier to establish familiarity with the
ICRC before a crisis than during a crisis when questions of impartiality may arise.

Support a Partnership with a Center of Excellence

USEUCOM and other unified commands should consider supporting
a partnership with a Center of Excellence (COE). A COE has benefits
for overall familiarization, information sharing, and long-term
planning.

As discussed in Chapter Eight, the COE assists coordination and
familiarization. It also provides institutional knowledge, which is
particularly important given the rapid turnover of military personnel.
At a minimum, this partnership might involve dedicating appropriate command assets to working with the COE. More ambitiously, it
might require the creation of a small agency analogous to the COE
but with a more restricted charter. USEUCOM has no need to dupli-
cate services already performed by existing COEs and generally
available to a wider community, such as training in disaster response
and data management. But USEUCOM could profit from a small
agency (approximately 6–8 people) dedicated to improving humanitarian response within the USEUCOM AOR. The agency might be
DoD-funded but responsive to a larger community of interested parties, including not only USEUCOM but also NATO and academic
institutions.

More fundamentally, USEUCOM and other unified commands
should promote development of DoD-level policy concerning agen-

cies like the COE. Currently, only USPACOM has easy access to such a center (USSOUTHCOM is in the process of establishing a similar center). If each unified command acts independently, some functions will be duplicated and some not accomplished at all, either for lack of sustained interest or lack of funding. Instead, there should be DoD-level policy to ensure that each unified command has easy access to a COE-like activity in its AOR and that all unified commands have access to one or more centers providing common functions.

As Table 11.1 suggests, the tasks associated with assuring NGO-military familiarity require the cooperation of a range of actors, including civilian agencies such as USAID. The armed services and the unified commands can take the lead in ensuring better familiarity, but our recommendations require the support of more than one institution.

IMPROVE INFORMATION SHARING: RECOMMENDED ACTIONS

The military should encourage initiatives to improve information sharing before and during crises. Such initiatives will enhance the military's and the U.S. government's awareness of likely problems and challenges in the relief operation and increase planning time through better information. Three such initiatives are:

- Identify NGOs with on-the-ground networks

- Minimize disruption caused by classification

- Share after-action reports and improve debriefings.

A more detailed suggested division of labor for implementing these steps is presented in Table 11.2 (pp. 156–157).

USEUCOM and other unified commands must distinguish those NGOs that have strong local networks from those that do not—a distinction that is often vital for judging the quality of information. Although NGO knowledge of local conditions varies from case to case, in general those NGOs with strong grassroots ties often are far better informed than those that do not. Frequently, religious NGOs, such as ADRA or Catholic Relief Services, have strong local networks as do those working on long-term development, such as CARE.[8] Generalizations are difficult, however, and it would be beneficial if USEUCOM and other commands knew which NGOs had a long-standing grassroots presence in countries in the AOR.

Although the intelligence community has met often with NGOs to share information, there is no policy on this relationship. Primarily for legal reasons, the community does not maintain a database on NGOs and their activities. Both NGOs and intelligence officials are also sensitive to any charges that NGOs have become intelligence sources. As a result, the intelligence community frequently does not know which NGOs are important, what information they possess, or how to access this information. The intelligence community must also disabuse relief personnel of the idea that it is omniscient during a crisis. As one intelligence official noted:

> Some outside the U.S. government think that just because the Government has so many resources devoted to information and intelligence collection and analysis, it MUST know almost every-thing about almost anything. In fact, that is not true. There are un-knowns. There are unknown unknowns. There are unknowables.[9]

Both before and during a crisis, classification concerns disrupt relationships with NGOs by making the information flow appear one-way and raising suspicions that the military or the U.S. government is deliberately concealing information. Another intelligence community member noted that intelligence agencies tend to remove far too much content from intelligence when sanitizing it and are often far too strict when classifying information. NGO personnel do not understand why some information is classified and resent being denied access. They particularly resent being confronted with access problems in a CMOC. The military should consider liberalizing its policy

---

[8]Catholic Relief Services, for example, has been active in Rwanda for 33 years and had a presence in Yugoslavia before World War II.

[9]Schoettle (1998).

Table 11.1

Ensuring NGO-Military Familiarity: Suggested Division of Responsibilities

Task	USAF	All Services (Title X Capacity)	Unified Command	Joint Staff	DoD	Other U.S. Government
Appoint an individual responsible for ensuring familiarity with relief community			Create a Humanitarian Affairs Advisor (HUMAD) position to liase with NGOs and IOs HUMAD will monitor activities of NGOs and IOs in countries within the AOR HUMAD will track command staff with relief experience When necessary, HUMAD will deploy with early arriving forces			
Brief relief community about military capabilities		Provide information on service capabilities in complex contingencies to NGOs and IOs through publications, liaison visits, and exercises	Inform NGOs and IOs of command relationships, capabilities, and planning	Provide information to NGOs through OFDA/ InterAction and to IOs through U.S. Mission to the UN		

Table 11.1—continued

Task	USAF	All Services (Title X Capacity)	Unified Command	Joint Staff	DoD	Other U.S. Government
Integrate civil affairs and SOF		Strengthen civil affairs and SOF liaison capability; ensure CA and SOF are aware of service-specific needs	Call on CA/SOF for regular briefings on relief agencies before crises begin		Expand civil affairs for peacetime operations	
Sponsor conferences and seminars	Sponsor conferences and seminars related to airlift	Sponsor conference and seminars related to service-specific capabilities	Sponsor conferences on crises in states in the AOR	Sponsor activities such as CJCS Peace Operations Seminar		
Develop and implement the Center of Excellence (COE) concept		Establish contact with the COE for training purposes	Support partnership with the COE Consider establishment of COE-like element in AOR	Establish guidance for COEs across unified commands	Develop policy for COEs; gain funding	USAID/OFDA: Assign personnel to COEs

NOTE: AOR = area of responsibility; CA = Civil Affairs; CJCS = Chairman, Joint Chiefs of Staff; COE = Center of Excellence; HUMAD = Humanitarian Affairs Advisor; IO = international organization; NGO = nongovernment organization; OFDA = Office of Foreign Disaster Assistance; SOF = Special Operations Forces; and UN = United Nations.

on classification to improve information sharing during crisis.[10] The military should announce classification guidelines in simple, direct language and classify only that information that would have direct military value to an opponent. It should routinely excise classified information from situation reports and share those reports with NGOs so that all interested agencies will share a common picture of the crisis. Important information to share includes safety, security, and medical information. In general, the U.S. government favors the dissemination of such information to aid agencies.

The military should share unclassified versions of its after-action reports with the United Nations, the ICRC, and NGOs. In return, it should expect to share other agencies' comparable reporting. Most larger operations generate a plethora of after-action reports and performance assessments. An experienced operator remarked, "If we could feed people with assessments, there would be no hungry people." But these reports often remain with the originators rather than being shared.[11] Sharing them would make the military and other actors more acutely aware of mutual problems. Similarly, if the military debriefed knowledgeable NGO personnel, it might improve overall engagement efforts.

When sharing information, however, the military must recognize that even information shared with core-team NGOs will not necessarily be closely held. NGOs in general do not appreciate the need for secrecy and regularly share information with all who will listen. At times, this information may go to partisan NGOs, local warlords, or hostile governments.

As Table 11.2 suggests, the tasks associated with improving information sharing—like the other tasks involved in improving military coordination with relief agencies—requires the cooperation of a range of actors, including several civilian agencies. The unified commands can take the lead in improving information sharing, but all the rec-

---

[10]Information sharing with NGOs may require a change in doctrine. Current doctrine notes that, "In the absence of sufficient guidance, command J-2s should share only information that is mission essential, affects lower-level operations, and is perishable." Joint Chiefs of Staff (1996), Joint Pub 3-08, p. III-21.

[11]Wentz (1998).

ommendations above require the support of more than one institution.

IMPROVE LONG-TERM PLANNING: RECOMMENDED ACTIONS

Beyond familiarization and information sharing, the unified commands should work with a small number of select NGOs to consider several steps to improve long-term planning and coordination. The small number reflects both the reality of the unified command's limited resources and recognition that the core NGOs do make the largest contributions to relief operations.

Such a selective approach will enable both the unified commands and the NGOs to work more closely before a crisis. These NGOs could help the commands establish better relations with the wider NGO community and serve as partners before trouble erupts. During a crisis, this improved relationship will help speed a deployment and make it more efficient.

The unified commands, the armed services in their Title X capacity, and other U.S. government actors should take these steps:

- Establish continuing contact with key NGOs

- Invite key NGOs into the planning process

- Develop relief packages

- Conduct more realistic exercises

- Consult with key NGOs about emerging crises

- Transport personnel from key NGOs.

A division of labor for implementing these steps is suggested in Table 11.3 (pp. 164–166).

Selection of key NGOs for closer engagement will help focus command efforts. Selection should not imply any discrimination against NGOs that are not selected. To preclude misunderstanding, the list of key NGOs should be informal and not disseminated. There should be no rigid selection criteria and the list should be open to constant revision.

Table 11.2

Improving Information Sharing: Suggested Division of Responsibilities

Task	USAF	All Services (Title X Capacity)	Unified Commands	Joint Staff	DoD	Other U.S. Government
Identify NGOs with good local networks			Identify NGOs willing to share information through the HUMAD and COE	Establish national headquarters-level contacts with core NGOs		USAID/OFDA: Inform unified commands of NGO activity in their AORs through InterAction and OCHA
Minimize disruption caused by secrecy			Minimize classification and make it intelligible to IOs and NGOs Prepare and disseminate unclassified after-action reports	Establish guidelines to minimize classification of data relevant to humanitarian assistance		

Table 11.2—continued

Task	USAF	All Services (Title X Capacity)	Unified Commands	Joint Staff	DoD	Other U.S. Government
Conduct debriefings and use after-action reports	Debrief TALCE personnel and others involved in air operations	Prepare and share unclassified after-action reports of operations	Conduct after-action conferences to review lessons learned	Make after-action reports available across unified commands and to NGOs at headquarters level	Contribute to after-action reports required by PDD-56	Commerce, Justice, NSC, State: Contribute to after-action reports required by PDD-56

NOTE: AOR = area of responsibility; COE = Center of Excellence; HUMAD = Humanitarian Affairs Advisor; IO = international organization; NGO = nongovernment organization; NSC = National Security Council; OCHA = Office for the Coordination of Humanitarian Affairs; OFDA = Office of Foreign Disaster Assistance; PDD = Presidential Decision Directive; and TALCE = Tanker Airlift Control Element.

The unified commands should experiment by first engaging a few candidates chosen from the list of core NGOs. Initial candidates professional relief organizations that can provide a variety of services and whose personnel have expressed willingness to work with the military. After gaining experience and overcoming any unexpected difficulties, the commands should expand the list until it includes all of the core NGOs. The command might work with USAID and with InterAction to choose the most appropriate NGOs for the AOR. Although working with all relevant NGOs, including specialized and minor organizations, has value, resource constraints will require that the unified command focus only on the most important and capable organizations.

Some NGOs may not want to be selectively engaged, particularly core-individual organizations. Growing NGO recognition of military contributions, however, and the benefits of ties to the military have made even some formerly hostile NGOs more receptive to better relations. The unified commands should encourage all the large and competent NGOs to participate, even while recognizing that during a crisis only those that have developed a solid ability to work with the unified command will receive preferential treatment. Even if many core NGOs choose to remain at arms distance from the military, closer contact with a few core NGOs will help improve unified command planning and relief capabilities.

Establish Continuing Contact

As part of selective engagement, the unified commands would establish appropriate continuing contact with key NGOs. These contacts might not form a consistent pattern. For example, in a highly centralized NGO, a single contact point might suffice, whereas in a less-centralized NGO, several contact points might be necessary to cover the USEUCOM AOR adequately. Both the COE and a HUMAD would be useful in helping the command establish continuing contact.

Establishing contact before a crisis is highly beneficial. As noted above, NGOs rely heavily on personal relationships and are less likely to work with the unified command if they do not know the people involved. More important, relationships forged during a crisis are far more likely to be seen locally as compromising impartiality. If the relationship is long-standing, however, NGOs can better claim that cooperation with U.S. forces is part of their normal routine rather than a response to a particular warlord's action or other threatening event.

Invite Key NGOs into the Planning Process

After laying the groundwork by establishing close contacts, the unified commands would invite key NGOs to participate in the planning process. Unified commands would invite NGOs to participate both broadly in a deliberate planning process and more explicitly in crisis planning. When planning begins for a joint task force, for example, key NGOs can provide useful information, help estimate relief requirements, and cooperate in providing relief packages.[12] At this stage, the key NGOs will want to hear how the unified command expects the operation to unfold. They will want to hear straightforward briefings on operational topics, which will demonstrate that the command wants to cooperate with NGOs and views cooperation as a two-way relationship.

Ideally, NGOs will change their procedures and activities to capitalize on command capabilities. NGOs will not accept tasking or formally designate responsibilities, but, if they believe cooperation with the command is in their interest, they will change their procedures accordingly. In a narrow sense, the unified commands cannot plan efforts of agencies that are not bound by their plans. But they can plan to support or accommodate these agencies' efforts on the assumption that they might participate. Thus, if the unified commands can improve the relief community's access to lift, communications, security, and other unified command assets, the agencies would be more likely to cooperate with the unified commands.

Develop Relief Packages

A logical third step would be to develop common understanding of relief packages that key NGOs could provide during a humanitarian crisis. With better coordination, the unified commands could help transport and distribute aid packages in the first few days of a crisis,

---

[12]Dworken (1996), p. 25.

when military assets may be the only ones available. NGO-provided packages might be designed to address particular needs, such as water purification, food, shelter, sanitation, immunization, or they might be fully rounded survival packages. At a minimum, measles vaccines, oral rehydration salts, and chlorine are highly useful in the early days of a crisis. Unified commands and key NGOs could then estimate the types and amounts of military or commercial lift that would be required to deliver the packages under various scenario assumptions.[13] The packages could then be integrated into planning contingencies. The unified commands should work with USAID and the Department of State to ensure adequate funding for these initiatives.

Some key NGOs have external quick-response capabilities, drawing upon expatriate personnel and prepositioned supplies. Others depend more heavily on internal capabilities, using indigenous personnel and local contracts. But even in this case, the NGOs may require assistance from the U.S. military during the initial phase of a crisis. The unified commands need to understand these varying capabilities and how assistance might be packaged to arrive most expeditiously.

Conduct More Useful Exercises

Many current exercises do not fully meet NGO or unified command needs and thus are less useful for long-term planning. Some exercises are not realistic regarding the role of relief agencies, and others take the cooperation of relief agencies for granted. Most NGOs, especially the core organizations, are busy responding to nearly continuous crises. NGO personnel usually schedule their time closely and resent wasting it. They are quick to sense when their participation is marginal or mere atmospherics.

When asked to participate in exercises, NGO personnel should be players whose inputs make a difference. In addition, they should be asked to help prepare the exercises or at least be consulted concern-

---

[13]Currently, OFDA is exploring an Indefinite Quantity Contract (IQC), which leads NGOs to specialize and prepare to meet a particular need. This, in turn, is leading many smaller NGOs to consolidate in order to receive U.S. funding.

ing appropriate roles for NGOs. To the extent possible, exercises should include free play that allows NGOs to act as they would in the field. Few NGO representatives will evince much interest in scripted play. Finally, exercises should not be designed to flow smoothly. They should raise difficult problems that have recurred in past operations such as chaotic airflow, the presence of refugees on a runway, or competing priorities for lift. Ideally, NGOs would also be brought into field exercises, as this is more likely to force them to demonstrate their flexibility and innovation, which are among their greatest assets during a real crisis. Raising such problems in exercises can help NGOs in particular to appreciate how uncoordinated efforts can make the entire operation less effective.

To get the most from UN and NGO players, unified commands should grant them major roles. The United Nations and NGOs, not the unified commands, will normally provide the bulk of humanitarian aid and nearly all of the interface with recipients. The unified commands support these other actors by responding to their requirements. They are not incidental to the operation; they are central to its very purpose. Therefore, an exercise should reflect their centrality and allow them to be as demanding and even obstreperous as they would be during an actual crisis. To obtain this effect, the unified commands should obtain, if possible, participation by NGOs that are less inclined to cooperate or more zealous in preserving their neutrality. Core-individualist NGOs such as MSF would be ideal participants. The whole point is to learn how the unified commands can support relief efforts by other actors, not how they might fit into unified commands' planning.

The unified commands should also consider paying the expenses that NGOs incur during exercises. Even the larger NGOs have limited budgets for activities outside their normal programs. Offering reimbursement would make participation easier for them. For specialized NGOs, financial assistance may be essential.

Consult with Key NGOs in Crisis Situations

During an emerging crisis, the unified commands should consult with key NGOs to obtain their views on impending humanitarian disasters and appropriate international responses. Such consultation would enrich the commands' understanding of the situation, help the commands recognize the requirements, and prepare for smoother execution of relief operations. Under condition of confidentiality, the commands might consult with key NGOs even before tasking from the national command authority (NCA) in order to better meet NCA directives. However, the commands would have to define their position clearly to avoid false expectations of support. When coordinating in advance, unified command officials must remember that relief agencies are often open with information, and shared information may not be handled discreetly.

Transport Personnel from Key NGOs

As an inducement to improve cooperation, the United States could offer to transport personnel from key NGOs during crises using military aircraft and other transportation assets. These personnel might include managers, sanitation experts, medical specialists, and others whose services were urgently required. Both U.S. military personnel and NGO officials noted that almost every other country's military was more able and willing to transfer personnel than the United States.[14]

The U.S. military should consider both increasing its transport of relief personnel in emergencies and transporting core-team personnel more frequently in other situations. In an urgent humanitarian crisis, the CINC can approve the transport of small numbers of urgently needed civilian personnel using military aircraft if no commercial aircraft are available. In nonemergency situations—but ones where commercial transport is not available—the command should work with DoD for exemptions needed to transport vital personnel. Before a crisis occurs, the command could also preclear with DoD a small group of NGO personnel for transport by military aircraft. In a crisis, these precleared individuals could more expeditiously be transported on military aircraft.

---

[14]Under the Denton Program, for example, only cargo can be transported by space available; people require dedicated flights. This restriction ensures that the Air Force does not compete with commercial carriers and also limits its liability.

As Table 11.3 suggests, improving long-term planning requires considerable support by both the unified commands and several U.S. government agencies, particularly the Department of State and Office of Foreign Disaster Assistance. The armed services would provide selective contributions related to service-specific concerns, but the primary burden would be on the unified command. U.S. government agencies and the Joint Staff would work closely with relief agencies at the headquarters level and provide guidance, respectively.

AVOID THE POTENTIAL PITFALLS OF SELECTIVE ENGAGEMENT

Being selective is a practical necessity, if only because the commands could not afford to cultivate relationships with hundreds of NGOs indiscriminately. However, there are potential pitfalls to selective engagement that the unified command and the military in general should recognize in advance and take care to avoid.

Allegations of Favoritism

Other NGOs might notice that key NGOs receive more attention and oppose selective engagement as a result. Some NGOs might acknowledge that greater capabilities understandably imply closer relations, while others might feel slighted.[15] In the latter case, the other NGOs might insist on equal treatment or even raise the issue with their donors and political constituencies.

Networking and inter-NGO relations are important to all NGOs, which means that peer opinion affects their willingness to cooperate with the military. Even core NGOs might hesitate to work closely with the military if other NGOs objected. NGOs are vulnerable to accusations of having "sold out"—being used as an instrument of U.S. policy rather than to serve humanity. The growth in NGO numbers and influence results from their ability to network, strategize, and

---

[15]NGOs are accustomed to being divided into "establishment" and "anti-establishment" groups. Although there are many rivalries and disputes among them, the NGOs in general have learned to use this diversity to good strategic effect. This will probably remain true even if a select few are "certified" and others are not.

Table 11.3

Improving Long-Term Planning: Suggested Division of Responsibilities

Task	USAF	All Services (Title X capacity)	Unified Command	Joint Staff	DoD	Other U.S. Government
Establish contact with core NGOs			HUMAD and COE-like element maintain continuing contacts; key officers (CINC, J-3, J-5,) component commanders have sporadic contacts at country level	J-5: Maintain continuing contacts with core NGOs at headquarters level		USAID/OFDA: Maintain continuing contacts at headquarters level
Invite core NGOs into the planning process	Invite core NGOs to participate in planning airlift		Invite NGOs to contribute to pre-planning and crisis response			

Table 11.3—continued

Task	USAF	All Services (Title X capacity)	Unified Command	Joint Staff	DoD	Other U.S. Government
Encourage NGOs to develop relief packages	Advise on airlift; develop mechanisms and procedures for prompt airlift of relief packages		Identify NGOs with rapid-response capabilities in AOR Plan command support of relief packages (both TRANSCOM and regional commands) Develop procedures and mechanisms for prompt delivery of relief packages	J-5: Provide guidance to unified commands for military support of relief packages		State and USAID: Host conferences of IOs, NGOs, and CINCs developing relief packages Fund relief package development
Conduct more useful exercises	Conduct exercises that focus on air-flow issues	Encourage IOs and NGOs to participate in planning on service-specific concerns Give IOs and NGOs substantive roles to play	Encourage IOs and NGOs to participate in planning Give IOs and NGOs substantive roles to play Conduct exercises that integrate military and civilian efforts	Contribute to PDD-56 exercises Draw unified commands into PDD-56 exercises	Promote, sponsor, and contribute to PDD-56 exercises	NSC and State: Promote, sponsor, and contribute to PDD-56 exercises

Table 11.3—continued

Task	USAF	All Services (Title X capacity)	Unified Command	Joint Staff	DoD	Other U.S. Government
Consult with core NGOs about emerging crises	Consult with core NGOs about airlift requirements and capabilities		Solicit views of NGOs at country level through HUMAD and the COE			USAID/OFDA: Solicit views of key NGOs at headquarters level
Transport NGO personnel during crisis			Maintain updated rosters of pre-approved NGOs personnel through HUMAD	J-4/J-5: Provide guidance to unified commands concerning transport of NGO personnel	Establish policy for transport of NGO personnel during crisis	USAID/OFDA: Request NGOs maintain updated rosters of key personnel eligible for transport by DoD

NOTE: COE = Center of Excellence; HUMAD = Humanitarian Affairs Advisor; IO = international organization; NGO = non-government organization; NSC = National Security Council; OFDA = Office of Foreign Disaster Assistance; PDD = Presidential Decision Directive; TRANSCOM = U.S. Transportation Command; and UN = United Nations.

divide tasks between them. As a recent study carefully documents, the NGO community has shown a remarkable aptitude for maintaining cohesion in the face of national and international efforts to drive wedges between them and to make clever tactical use of their differences.[16] For example, "establishment" NGOs who gain admission to official meetings will generally be scrupulous about holding briefings, strategy sessions, and the like with those NGOs who failed to make the cut. The better established ones lobby, sit on UN committees, and help draft resolutions, but they know that at least part of their weight comes from the fact that other NGOs are in front of the building with placards, demonstrating and issuing protest statements about a current policy.

To prevent charges of "selling out" from arising, the unified commands should keep selective engagement informal and flexible. It should treat smaller NGOs with respect and keep them well informed of command initiatives that could affect relief operations. It should also stress that attention is given strictly because of an NGO's overall capabilities and ability to work with the U.S. military: If smaller NGOs develop these traits, then they too will receive closer attention.

Concerns Regarding Independence

Key NGOs would avoid closer relationships with unified commands if they feared that their independence could be compromised. Although most regularly receive U.S. government funding, they rightly insist on the neutrality and impartiality implicit in their humanitarian charters. Quite apart from moral considerations, they arguably would be less useful to the U.S. government if they were not independent. The unified commands can avoid raising such concerns if they recognize two principles: (1) in relief operations, the military normally supports NGOs, not the other way around, and (2) relations between the military and NGOs are voluntary and cooperative. During actual deployments, U.S. forces must also recognize that NGOs may vacillate in their willingness to associate with the military and that preserving NGO impartiality is likely to facilitate overall success.

---

[16]The growth of NGO influence as a result of a determined networking, planning, and strategic effort is well documented in Clark, Friedman, and Hochstetler (1998).

Cross-Purposes

The relationship of military authorities with NGOs is usually medi-
ated by U.S. government agencies except during actual operations
when direct contact, for example through a CMOC, becomes essential. Often, OFDA or a country team works directly with NGOs while
the military responds to tasking. By bringing the military into direct
contact with NGOs, selective engagement risks leading the military
to work at cross-purposes with other government agencies. For example, USAID might prefer one of its traditional U.S.-based NGO
partners for a particular task, whereas USEUCOM might prefer an
NGO based in Europe.

To prevent disconnects of this sort, the unified commands should
keep relevant U.S. government agencies informed of its precrisis engagement and during crisis it should work closely with them. Close
coordination with U.S. government agencies that also work with
NGOs will be necessary in any event to ensure the success of selective engagement. OFDA can encourage elements of selective engagement, such as relief packages, with financial support. In general,
NGOs will be more inclined to cooperate with the military if they
realize that their ties to the U.S. government will improve if they do.

Strain on NGOs

A demanding engagement strategy might put too much strain on key
NGOs. In interviews, several large NGOs noted that they could not
afford to provide personnel to attend all activities sponsored by the
military. From their perspective, the military is a gigantic organization that can easily overwhelm their slender personnel resources. To
avoid putting too much strain on key NGOs, the unified command
should make contacts brief and intense with little wasted time. It
should also send its officers to the key NGOs rather than always
having NGO personnel come to them.

Unfounded Expectations

Unless carefully managed, selective engagement could raise unfounded expectations among key NGOs. Past military support for
relief operations has often been episodic, unpredictable, and driven

by political motives. Selective engagement could convey an impression that the U.S. government is initiating a new policy of broader and steadier support but then disappoint NGOs if the U.S. government chooses not to intervene in a particular crisis. To avoid raising unfounded expectations, the unified commands should make certain key NGOs understand that large-scale military support is contingent upon NCA tasking case-by-case.

IMPROVE THE COORDINATION OF THE RELIEF FLOW: RECOMMENDED ACTIONS

The regional commands and USTRANSCOM can use their tremendous logistics capabilities to improve the overall flow of relief goods to a crisis region. Particularly in the early days of a crisis, the flow of relief is chaotic and sporadic, which can lead to shortages of critical goods, delays, and other problems. In general, ground transportation presents few problems for NGOs and IOs. Airlift, and to a lesser extent, sealift, is a far more complex problem, and NGOs lack the ability to manage large relief efforts that involve these forms of transportation.

Poor coordination, approaching chaos at times, is a recurring problem in humanitarian airflow. During Operation Support Hope, there was near chaos at receiving airports. In some instances, civilian aircraft chartered by NGOs simply appeared unannounced and had to be diverted because of congestion. Initially, there was little overall prioritization of relief efforts, so that unneeded items were as likely to arrive as desperately needed items. Rwanda is a particularly striking and dramatic example, but similar lack of coordination afflicts airflow during nearly every large humanitarian operation.

Better coordination of the relief flow requires several interrelated tasks that necessarily involve a large number of actors including host countries, donors, the United Nations, and NGOs. The military as a whole and unified commands in particular have only limited influence over some of these actors, but they can promote workable solutions. Success requires a strong effort by other U.S. government agencies, to which the services and Department of Defense could contribute. Fundamental tasks include:

- Set overall priorities for the relief effort
- Ensure adherence to a common schedule
- Provide logistics management control and off-loading.

A more detailed division of labor for implementing these steps is suggested in Table 11.4 (pp. 176–177).

Set Priorities for Relief Effort

The first and most important step is to set priorities for relief efforts based on a common understanding of the amounts and types of aid that are required over time. The military cannot set these priorities but it needs them to work efficiently. As noted earlier, the structure that sets priorities may be characterized as host country, United Nations, alliance, or coalition.

If a host country maintains governance, it may set priorities or it may simply welcome any assistance that arrives. In such cases, the military usually operates in mixed-use facilities, sharing port facilities, ramp space, and slot times with civilian organizations. Host country authorities may willingly cede de facto control or executive agency status to the military when they perceive that it can operate airports most efficiently. Often, the host country may offer only partial use of an airfield for the relief effort.

If the United Nations takes the lead with the support of the U.S. government, the unified commands should coordinate closely with offices of its key agencies in Geneva and with their representatives in the field. Through its own actions, the unified commands should support whatever option the United Nations has chosen to coordinate its response, whether through an Emergency Response Coordinator or through a lead agency such as UNHCR.

Within the USEUCOM AOR, for example, NATO might assume control, especially for relief efforts in the Balkans. The entire Alliance might act pursuant to decisions taken in the North Atlantic Council (NAC), or a coalition of willing members might use Alliance resources. In either case, the Alliance would have to set priorities in cooperation with agencies of the United Nations, which would also be involved. Almost certainly, Alliance members would also be the largest donor states.

An individual country might lead others, as the United States did during Operation Provide Comfort. Within the USEUCOM AOR, the lead state might be the United States, France, or possibly Italy, as during Operation Alba. This lead state would set priorities in cooperation with other interested states, whether participants in the operations or merely donors, and with UN agencies. If a foreign country were to lead, USEUCOM (or other unified command) would have to establish liaison with its military authorities. Once an operation is under way, USEUCOM can establish a CMOC. It should then become a forum to reach agreement on priorities among relief providers.

Both the regional unified commands and USTRANSCOM (particularly AMC) must ensure that goods moved under space-available flight provisions follow relief priorities. Currently, items shipped under space available are not prioritized: What is shipped first depends on local flight availability and chance. The goal of planners is to maximize what is sent, not to ensure that what is sent is needed immediately. The unified commands and USTRANSCOM should explore ways to prioritize space-available cargo when possible.

Ensure Adherence to a Schedule

Once priorities are set, the next concern is to ensure adherence to a common schedule. Aid often arrives haphazardly and chokes small ports or airfields. Especially in the early days of a crisis, NGOs and UN agencies are not able to manage the complex and massive aid flows, particularly if they involve airlift.

Schedule problems are particularly acute for effective airlift. All relief agencies that conduct or sponsor flights into the affected region must accept their places in the aid queue and plan flights accordingly. They must conform to appropriate procedures regarding slot times and other crucial aspects of the operations such as allocation of ramp space.

Usually the military will lack authority to ensure complete adherence to schedules, but it can work through the coordination structure to encourage adherence. It can emphasize to host countries that maintaining a proper schedule will ultimately raise the level of humanitarian assistance, even if some flights must be turned away. It can work through UN agencies and ultimately through donor countries to ensure that all NGOs are kept informed on procedures. Most of the larger NGOs receive substantial funding through governmental channels and through UN agencies and are anxious to impress these sponsors with their reliability and professionalism. Although NGOs might complain about restrictions on movement, interviews suggest they would comply as fully as possible with procedures imposed by large donors. Many NGOs also recognize the problems that come with anarchic flow of aid and thus are more willing to cooperate.

Admittedly, perfect adherence to schedules will seldom be possible, even when an individual state leads the operation. There will almost always be donor nations that act unilaterally, UN agencies that fail to coordinate perfectly, and NGOs (especially smaller, less professional ones) that send or sponsor flights without reference to schedules. In some instances, lack of compliance may be willful, reflecting political decisions or rivalry among agencies. In other instances, lack of compliance may be inadvertent, reflecting inexperience or innocent zeal.

Provide Logistics Management

USTRANSCOM and the service staffs can support the regional unified command in an effort to help manage the flow of aid. NGOs and UN agencies in general are experienced with transporting goods by land. Moreover, they often have in-place networks that have been delivering aid long before the NCA decided to act. For aid delivered by sea, however, USTRANSCOM can help coordinate the flow and improve port assets. This logistics assistance should be provided until UN agencies and NGOs have the personnel in place to assume responsibility for the effort.

Improving airlift should be a key part of the logistics effort; this capability is weak among NGOs and is often important in sudden and massive crises, in which relief agencies may be overwhelmed. USAF elements in USTRANSCOM can perform air traffic control and other functions essential to efficient airport operations. AMC has Tanker

Airlift Control Elements (TALCE)—and other air traffic management assets that contribute to an Air Mobility Element (AME) for larger operations—constantly on alert to support airlift in the context of military operations. The capability embodied by TALCE is virtually unique to the United States. Among the NATO allies, only France has a comparable capability to maintain her overseas commitments. It is not clear whether NATO itself could perform as well. WFP can manage only smaller airlifts within its own programs. UNHCR has no capability to direct an airflow, unless it is augmented as it was during the Rwanda operation. Therefore, in a humanitarian crisis that requires rapid deployment, there may be no practical alternative to U.S. TALCE. Over time, when airlift becomes more routine, the TALCE can hand off its activities to the host nation, the United Nations, or other body.

USTRANSCOM, USEUCOM, and other commands could prepare for deployment of TALCE in support of relief operations in several ways. They could conduct training and exercises that include coordination of civilian aircraft in scenarios involving humanitarian aid. (TALCEs normally control only U.S. military airlift and control civilian aircraft by exception.) They could acquaint UN agencies and NGOs with the capabilities of TALCEs. Without committing the United States to any particular course of action, they could explore with OCHA and UNHCR how TALCEs might be used. USEUCOM's 86th Contingency Response Group (CRG), set up to rapidly deploy and run an airfield, performed well during Operation Shining Hope and should be emulated by other commands. Augmenting the CRG with personnel familiar with NGOs would make it even more effective.

Air traffic control assets should be employed as early as possible in a crisis, subject of course to diplomatic realities. Often, a TALCE is called in only after a problem develops, and at times is not deployed until weeks into a crisis. An AME may not be used at all. Because the early days of a crisis are often the most deadly and chaotic, employing this capacity earlier—even if at times it is not absolutely necessary—could help the relief effort considerably. Once employed, a Temporary Flight Restriction could be issued, to let all carriers know they must have prior permission to land from the air traffic control element, acting on behalf of the host nation or the lead country or agency.[17]

USTRANSCOM, USEUCOM, and other commands could also prepare to receive foreign personnel, including personnel drawn from UN agencies and NATO militaries, into TALCEs during relief operations. Foreign personnel could provide expertise from their national forces. In addition, their presence would give a TALCE an international flavor that would make donor nations less reluctant to accept control than if the TALCE were exclusively composed of U.S. personnel. But to ensure efficiency, foreign personnel should augment a TALCE, not occupy key positions. It would be impractical to assemble a truly international TALCE during crisis.

As Table 11.4 outlines, most of the burden for ensuring a smoother flow of aid will fall upon the service components, particularly the USAF, and the unified commands. The services will provide the capability, and both the services and the commands will ensure that adequate procedures exist for relief agencies to use the capability. A smoother aid flow will also require the effective transmission of priorities to the relevant military officials. In the early days of a crisis, this will be done primarily by the lead nation or agency; over time, as the CMOC is established, operational priorities will be generated locally, with U.S. government agencies providing political input.

ESTABLISH INITIATIVES WITH ALLIES

Relief operations are increasingly multinational and complex, with ever-increasing interaction between civilian actors and military establishments. European allies are a leading part of this equation, worldwide. European NGOs are among the most active in humani-

---

[17]In a crisis, TALCEs might be organized in two different ways. Broadly speaking, air operations might be centralized or decentralized, depending upon the exigencies of the situation. During the Rwanda operation, UNHCR received augmentation from USEUCOM and attempted to exert centralized control from Geneva, analogous to an arrangement made to control airlift in Bosnia. In Bosnia, this arrangement was appropriate because a single airport (Sarajevo) dominated traffic. But this arrangement was inappropriate in Rwanda because the operation involved several destination and staging airports whose operations could not be efficiently controlled from Geneva. Moreover, Geneva was less well informed of the rapidly developing situation than were elements on the ground in Rwanda and Zaire.

tarian relief, and the European Union itself is emerging as the largest humanitarian actor in key regions. In the wake of the Kosovo crisis, NATO's role in relief operations in the Euro-Atlantic area has become more prominent, raising new issues of coordination and civil-military relations. This analysis suggests a number of key findings and points to areas for new initiatives at the strategic and operational levels.

U.S. and USAF policy should aim at capturing useful synergies with European allies. At the strategic level, decisionmakers can take advantage of existing European defense relationships, facilities, and experience, especially in Africa. Similarly, the U.S. comparative advantage in technical intelligence on regional developments can be reinforced by European strengths in intelligence collection on the ground. At the operational level, U.S. capacity for strategic airlift complements the European capacity for tactical lift that most humanitarian contingencies require. Many of these recommendations will require broad support from civilian agencies of the U.S. government.

Strategic and Political Initiatives

Given the growing role of European allies and changes in the involvement of key organizations, there are now worthwhile opportunities to improve cooperation with allies at the strategic level.

- *Strengthen NATO's capacity for civil emergency planning and humanitarian relief.* Building on the experience in IFOR/SFOR and KFOR and as part of the implementation of NATO's new Strategic Concept, the relevant organizations within NATO (especially the EADRCC) should be strengthened, particularly by providing the necessary resources for training and exercises with partner countries. The profile of civil emergency planning might be raised through the establishment of a NATO Assistant Secretary General (ASG) for Civil Emergency Planning. Among other responsibilities, a NATO ASG for Civil Emergency Planning could facilitate Alliance planning and coordination with key IOs and NGOs. The Kosovo experience should spur interest in high-level dialogue between the Alliance and UN organizations.

Table 11.4

Ensuring a Smoother Relief Flow: Suggested Division of Responsibilities

Task	USAF	All Services (Title X Capacity)	Unified Command	Joint Staff	DoD	Other U.S. Government
Set overall priorities for the relief effort	Prioritize space available in support of USTRANSCOM		Perform command assessment of requirements for humanitarian relief			Before CMOC established, pass priorities down in consultation with on-the-scene NGOs
			Establish CMOC with direct connection to Operations Center and Air Cell			
			Encourage IOs and NGOs to develop local priorities within CMOC			After CMOC established, pass political priorities to Air Cell and other logistics cells

Table 11.4—continued

Task	USAF	All Services (Title X Capacity)	Unified Command	Joint Staff	DoD	Other U.S. Government
Ensure adherence to common procedures	Prepare for foreign personnel assisting a TALCE and directly participating in its operations		Establish and promulgate procedures through HOC and CMOC Establish procedures and pre-CMOC interface to ensure smooth delivery of early arriving goods and personnel (USTRANSCOM) Establish interface with relief agencies to provide cargo data, schedule slot times, etc. (USTRANSCOM and regional commands)	Develop and promulgate joint doctrine on common airlift procedures		USAID/OFDA: Require core NGOs to adhere to common airlift procedures Encourage core NGOs to develop common coordination procedures for early arriving relief
Provide early cargo traffic control and off-loading	Prepare TALCE and other necessary assets (comparable to 86th CRG) to manage humanitarian lift		Use logistics management capabilities to assist humanitarian relief effort Establish operational links to USTRANSCOM Ensure rapidly deployed forces have staff with NGO expertise available			

NOTE: CMOC = civil-military operations center; CRG = Contingency Response Group; HOC = Humanitarian Operations Center; IO = international organization; NGO = nongovernment organization; OFDA = Office of Foreign Disaster Assistance; TALCE = Tanker Airlift Control Element; USTRANSCOM = U.S. Transportation Command.

- *Put transatlantic cooperation in humanitarian crises high on the prospective EU-NATO agenda.* EU-NATO consultations will be a necessity as the EU's common foreign and security policy evolves. Many of the contingencies in which NATO (especially U.S.) assets may be placed at the disposal of future European-led operations are likely to be humanitarian in nature. Humanitarian early warning and contingency planning should be key—and relatively uncontroversial—agenda items for EU-NATO dialogue.

- *Engage European allies in multilateral activities to bolster local capacity for humanitarian and peace support operations.* The United States and the EU, as well as key allies such as France and Britain, have made this approach a focal point of their regional security strategies. Multilateral exercises on the pattern of those already conducted in Africa should be continued and might usefully be extended to regions such as the Caucasus, Central Asia, and the Pacific.

Operational Initiatives

This analysis makes a number of recommendations for improved cooperation with allies, and through allied institutions, at the operational level.

- *Explore arrangements to take advantage of French facilities and European relationships in and around Africa to support relief operations.* French opinion is sensitive to U.S. policy and presence in Africa. But within limits, the humanitarian context may be one in which more formal access arrangements are possible. Expanded military-to-military cooperation (e.g., between USAFE and the French airlift command) may be the best vehicle for this. Even more important, training and exercises with European militaries, where possible in conjunction with local militaries, can contribute to local knowledge and working relationships in advance of future expeditionary deployments.

- *Promote interoperability and standardization in airlift/airdrop with European allies.* Key European militaries are interested in this objective, and given the large role of NGOs in this arena through commercial charters, cooperation could be extended to civilian actors, where appropriate. Particularly important are steps to enhance traffic management for military and commercial airlift.

- *Provide NGO spaces at relevant courses and war colleges.* NATO has invited representation from UN organizations at the NATO School at Oberammergau and the NATO Defense College in Rome. U.S. and European NGO representatives could also be included in training in relevant areas such as civil-emergency planning, logistics, force protection, and civil-military relations.

Initiatives along these lines can help to advance the level of cooperation with European allies in an area where Europe has a relatively full capacity for burdensharing. Military support to relief operations outside the NATO area is a sphere in which Europe already plays a leading role. In terms of overall humanitarian assistance, the EU is itself a leading actor—and this role is set to increase. In operational terms, there is significant "value added" to be gained from a closer operational relationship with allies given the European networks in Africa and elsewhere. These networks can be valuable in helping to anticipate and prepare for complex relief operations in an expeditionary environment.

Finally, the prospect of a greater NATO role in managing humanitarian crises through the Alliance's civil emergency planning structures and as a result of changing missions will benefit the United States and the USAF. In most cases, a NATO frame will facilitate working with allies and will help to institutionalize patterns of coordination with NGOs and international organizations.

FINAL WORDS

There are no complete solutions to the operational and coordination problems discussed in this study. Many of the solutions to these problems lie outside the USAF's, and the broader U.S. military's, spheres of responsibility. Many fixes dictate actions across the U.S. government, requiring the services, the unified commands, the Joint Staff, the Department of Defense, and civilian agencies to work closely together. Coordination on complex emergencies within relevant agencies of the U.S. government, however, is often poor, making problems that affect military performance difficult to solve.[18] Moreover, as the analysis in this study makes clear, U.S. institutions must work with UN agencies and NGOs, which have their own limits.

At a more fundamental level, the United States has not decided whether intervention in complex emergencies will be a central task for its military or a collateral one in the coming decades. Until that decision is made, the resources necessary to organize, train, and equip U.S. forces for interventions in crises, and the associated doctrinal developments, are likely to be lacking. Civilian agencies may not take the appropriate steps to improve their coordination with the military until this decision is made.

By keeping in mind the likely resource limits and policy constraints that stem from this indecision, military planners can help reduce overly optimistic expectations about what relief operations can accomplish and anticipate likely operational problems. Equally important, the military can improve coordination with relief agencies and with U.S. allies, thus avoiding some of these problems and minimizing others. The recommendations suggested in this chapter would make future operations go more smoothly, with fewer disruptions that can exacerbate the suffering of victims of humanitarian crises.

---

[18]Pirnie (1998).

U.S. NGOs

Appendix A lists U.S. NGOs by name, category, mission, geographic coverage, total 1995 revenue, dollar amount of U.S. government grant or contract, and dollar amount of U.S. government in-kind payments.

Appendix A

U.S. NGOs

Name	Category[a]	Mission	Geographic Coverage	Total Revenue in 1995 ($)	U.S. Govt Grant/ Contract ($)	U.S. Govt In-Kind ($)
ACDI/VOCA	Minor	Long-term technical assistance	Worldwide	23,214,432	13,703,048	9,341,307
Action Against Hunger	Specialized	Emergency relief and development	Worldwide	5,118,713	3,788,332	0
Adventist Development and Relief Agency International (ADRA)	Core-team	Immediate relief and development	Worldwide	59,515,100	11,681,820	35,975,400
African American Institute (AAI)	Specialized	Promote understanding	Africa	35,847,521	33,742,488	0
Africare	Core-team	Emergency relief and development	Africa	35,994,038	19,810,979	0
American Friends Service Committee (AFSC)	Minor	Longer-term projects	Worldwide	27,516,453	192,207	0
American Jewish Joint Distribution Committee	Core-team	Rescue, relief, reconstruction	Worldwide	81,272,103	0	0
American Red Cross International Services Dept	Core-team	Disaster response	Worldwide	35,396,888	7,182,762	5,778,780
American Refugee Committee (ARC)	Specialized	Relief and training of refugees	Africa, Asia, Europe	11,344,780	6,449,618	0
Baptist World Alliance	Minor	Disaster relief and rehabilitation	Worldwide	9,770,294	0	0

Name	Category[a]	Mission	Geographic Coverage	Total Revenue in 1995 ($)	U.S. Govt Grant/ Contract ($)	U.S. Govt In-Kind ($)
The Brother's Brother Foundation	Specialized	Distribution of in-kind contribution	Worldwide	50,518,525	393,511	0
CARE	Core-team	Emergency assistance and development	Worldwide	454,990,000	92,224,000	196,159,000
Catholic Medical Mission Board	Specialized	Emergency and longer-term health care	Worldwide	35,783,340	0	0
Catholic Relief Services	Core-team	Relief and development	Worldwide	270,118,000	67,139,000	127,647,000
Childreach	Specialized	Sponsor needy children	Worldwide	29,969,345	1,112,525	0
Children's Survival Fund	Minor	Relief and development	Worldwide	6,597,370	0	0
Christian Children's Fund (CCF)	Specialized	Development	Worldwide	111,423,662	717,481	735,409
Christian Reformed World Relief Committee	Minor	Relief and development	Worldwide	8,017,369	833,685	0
Church World Service (CWS)	Specialized	Relief and development	Worldwide	44,786,000	13,659,000	0
Counterpart International	Minor	Relief and development	Asia	16,621,752	5,012,150	0
Direct Relief International	Specialized	Medical relief	Worldwide	45,469,678	111,172	0
Doctors Without Borders USA	Core-individual	Medical relief	Worldwide	6,193,311	0	0
Food for the Hungry International	Specialized	Relief	Worldwide	22,242,736	7,312,269	0
Heifer Project International	Specialized	Agricultural development	Worldwide	10,474,154	928,526	0
Immigration and Refugee Services of America	Advocacy	Professional services and placement	Croatia and Rwanda	10,037,002	7,400,195	23,333
International Aid	Core-team	Relief and medical aid	Worldwide	45,687,711	0	0

Name	Category[a]	Mission	Geographic Coverage	Total Revenue in 1995 ($)	U.S. Govt Grant/ Contract ($)	U.S. Govt In-Kind ($)
International Catholic Migration Commission	—	Coordinates assistance to refugees	Worldwide	14,184,000	11,939,000	0
International Medical Corps	Specialized	Medical relief	Europe, Africa	11,126,869	7,999,163	0
International Orthodox Christian Charities	Minor	Relief and development	Europe	7,783,855	264,179	0
International Rescue Committee	Core-team	Relief and development	Africa, Asia, Europe	82,980,020	46,262,054	0
Lutheran World Relief (LWR)	Minor	Relief and development	Worldwide	21,793,433	1,727,436	0
MAP International	Specialized	Medical relief and assistance	Worldwide	111,275,423	0	224,305
Medical Care Development (MCD)	Specialized	Medical relief and assistance	Africa	7,529,021	2,526,777	0
Mercy Corps International (MCI)	Core-team	Development	Worldwide	34,724,855	6,532,463	16,331,762
Operation USA	Minor	Disaster relief supplies	Worldwide	7,663,830	64,156	6,994,289
Oxfam USA	Core-individual	Disaster relief and self-help	Worldwide	13,408,623	0	0
Partners of the Americas (POA)	Minor	Development	Americas	7,625,909	6,220,774	0
Relief International	Minor	Relief and development	Asia	7,004,510	1,557,137	625,000
Salvation Army World Service Office	Minor	Relief and technical assistance	Worldwide	23,100,823	698,142	10,461,468
Save the Children USA	Core	Relief and development	Worldwide	109,492,000	54,911,000	8,492,000

Name	Category[a]	Mission	Geographic Coverage	Total Revenue in 1995 ($)	U.S. Govt Grant/ Contract ($)	U.S. Govt In-Kind ($)
United Methodist Committee on Relief	Core-team	Relief and development	Worldwide	36,585,853	0	12,040,600
Volunteers in Technical Assistance	Minor	Information and communication	Emphasis on Africa	5,409,877	3,694,016	55,264
Winrock International	Specialized	Development	Worldwide	31,950,000	0	22,797,000
World Relief Corporation	Specialized	Relief and development	Worldwide	24,130,220	13,970,839	0
World Share	Specialized	Development	Guatemala and Mexico	45,545,762	2,427,695	4,104,147
World Vision Relief and Development	Core-team	Relief and development	Worldwide	320,047,000	47,926,000	0
YMCA USA	Specialized	Development	Worldwide	45,832,732	2,291,250	0
YWCA USA	Minor	Development	Worldwide	12,293,277	572,338	0

[a]Core-team highly competent, broadly capable, and predisposed to cooperate with the military.

Core-individual: highly competent and broadly capable, but less eager to cooperate with the military.

Specialized: highly competent and capable in functional areas.

Advocacy: focused on human rights on an agenda not necessarily related to providing immediate humanitarian relief.

Minor: competent but having less capability than the core-team.

MAJOR INTERNATIONAL NGOs

Appendix B lists major international NGOs by name, mission, structure, geographic coverage, and revenue and government support.

Appendix B
Major International NGOs

Name	Mission	Structure	Geographic Coverage	Revenue and Government Support
Action Against Hunger	"Nutrition, food security, water and sanitation, health programs, and disaster preparedness"; intervenes "where survival depends on humanitarian aid, when natural or man-made crises threaten food security or result in famine, where societies in upheaval render populations extremely vulnerable."a	Four branches: U.S., France (Action Contre la Faim), Spain (Accion Contra el Hambre), UK; regional office in Nairobi	Afghanistan, Angola, Armenia, Bosnia, Burundi, Cambodia, Cameroon, Chad, Colombia, Dem. Republic of Congo, Ethiopia, Ivory Coast, Georgia/Abkhazia, Guatemala, Guinea, Haiti, Laos, Liberia, Mali, Mozambique, Myanmar (Burma), Nicaragua, Niger, North Korea, Sierra Leone, Somalia/Somaliland, Sri Lanka, Sudan, Uganda	Individual donors; foundations; companies; UNHCR; UNDP; UNICEF; WFP; British, Norwegian, Dutch Cooperation Agencies; government support from USAID, EU (ECHO, European Community Humanitarian Office); French Ministries of Cooperation and Foreign Affairs, Dept. of Humanitarian Action

Name	Mission	Structure	Geographic Coverage	Revenue and Government Support
Adventist Development and Relief Agency (ADRA)	Five core areas: Economic development, food security, primary health, disaster response and preparedness (assist victims of natural and human-made disaster, short- and long-term assistance to populations of refugees and displaced people. Includes health care, water, clothing, shelter and housing reconstruction), and basic education.[b]	ADRA Japan, ADRA Netherlands, ADRA Norway, ADRA Sweden,	Headquarters in Silver Spring, MD (North American region); Regional offices in Cote d'Ivoire (Africa Indian Ocean region), Zimbabwe (Eastern Africa region), Switzerland (Euro Africa region), Russia (Euro Asia region), Singapore (Southern Asia-Pacific region), Florida (Inter American region), Brazil (South American region), Australia (South Pacific region), India (South Asia region), England (Trans European region)	FY 1997: Support from U.S. government/ USAID, including commodities, excess property, grants, ocean and inland freight ($44,957,173); cash donations ($7,093,969); donated materials and services ($6,620,738); General Conference of Seventh Day Adventists ($3,381,918); interest/other income ($1,352,286); ADRA funding ($688,158)
Cooperative for Assistance and Relief Everywhere (CARE International)	"To relieve human suffering, to provide economic opportunity, to build sustained capacity for self-help, and to affirm the ties of human beings everywhere." Focuses on basic education, stable food and water supplies, basic health care, access to family planning, safe and sustainable environment.[c]	Australia, Austria, Canada, Denmark, France, Germany, Japan, Norway, UK, U.S.		

Name	Mission	Structure	Geographic Coverage	Revenue and Government Support
Caritas Internationalis (Caritas Federation)	Priority actions, 1995–99: Causes of poverty/injustice; promotion of reconciliation / nonviolence, conflict management, etc.; migrants, refugees, displaced persons; responses to emergency situations.	160 members, organized by departments and regional desks		
Concern	Emergency response: "to crises such as natural disasters . . . and political or man-made disasters—usually civil war"; working in development against poverty, working in development education and adequacy.[e]	Concern Republic of Ireland (main office), Concern Northern Ireland, Concern Glasgow, Concern USA, Concern United Kingdom	Emergency programs in Sudan, Afghanistan, North Korea, Rwanda, Sierra Leone, Liberia, Somalia, Democratic Republic of Congo, Burundi, Honduras	FY 1997: Co-funding international governments: £10,510,000 (48%; Irish & British governments, EU); public donations: £8,190,000 (38%); donated goods and services, £2,634,000 (12%); deposit interest £387,000 (2%)

Name	Mission	Structure	Geographic Coverage	Revenue and Government Support
Doctors Without Borders/ Médecins sans Frontières	Aid to "victims of armed conflicts, epidemics, and natural and man-made disasters: others who lack health care due to geographic remoteness or ethnic marginalization . . . teams provide primary health care, perform surgery, vaccinate children, rehabilitate hospitals, operate emergency nutrition and sanitation programs, and train local medical staff."f	MSF International (Belgium), Australia, Austria, Belgium, Canada, Denmark, France, Germany, Greece, Holland, Hong Kong, Italy, Japan, Luxembourg, Norway, Spain, Sweden, Switzerland, United Arab Emirates, United Kingdom, U.S.	Central Africa: Burundi, Central African Republic, Chad, Democratic Republic of Congo, Rwanda. Eastern Africa: Ethiopia, Kenya, Madagascar, Somalia, Sudan, Tanzania, Uganda. Western Africa: Benin, Burkina Faso, Equatorial Guinea, Cote d'Ivoire, Guinea, Liberia, Mali, Mauritania, Nigeria, Sierra Leone. Southern Africa: Angola, Malawi, Mozambique. Europe: Armenia, Azerbaijan, Belgium, Bosnia-Herzegovina, France, Federal Yugoslav Republic, Georgia, Italy, Romania, Spain, Russia. Central and South America: Brazil, Bolivia, Colombia, Cuba, Costa Rica, Ecuador, Guatemala, Haiti, Honduras, Mexico, Nicaragua, Panama, Peru. Central and East Asia: Afghanistan, Bangladesh, Cambodia, China, Indonesia, Laos, Kazakhstan, Kyrghyzstan Myanmar (Burma), Philippines, North Korea, Sri Lanka, Tajikistan, Thailand, Vietnam, Uzbekistan. Middle East and North Africa: Egypt, Iran, Lebanon, UAE, Yemen, Palestinian Authority	FY 1997: $231 million; 54% from private donors; 18.25% from ECHO; 5.26% "other EU funding"; 4.63% UNHCR; 3.6% the Dutch government; 2.61% the Norwegian government; 2.19% the Belgian government; 2.09% the U.S. government; 1.76% the Luxembourg government

Name	Mission	Structure	Geographic Coverage	Revenue and Government Support
International Committee of the Red Cross (ICRC)	"What we do: visiting people deprived of their freedom, protection of the civilian population, war and family ties, relief operations, health activities: general information, dissemination and preventative action, humanitarian diplomacy, legal work, Advisory Service on International Humanitarian Law."g	Regional delegations (also individual country delegations), Cameroon, République de Cote d'Ivoire, Nigeria, Senegal, South Africa, Zimbabwe, India, Indonesia, Philippines, Thailand, Uzbekistan, Russian Federation, Ukraine, Argentina, Brazil, U.S., Guatemala, Kuwait, Tunisia, UN delegation in New York		Percentage of contributions received, in cash, kind, and services, 1997: 72.5% from governments, 14.5% from the European Commission, 9.5% from National Societies, 2.8% from private sources, 0.6% from public sources, 0.1% from international organizations; government contributions come from states party to the Geneva conventions
Mercy Corps International	"To alleviate suffering, poverty, and oppression by helping people build secure, productive, and just communities."h Provides emergency relief services for people afflicted by conflict or disaster; invests in community development projects; runs "civil society initiatives" that promote citizen participation.	Headquarters in Oregon; regional offices in Washington, DC and Washington State; Pax World Service (Washington, DC); Mercy Corps Europe/ Scottish European Aid (UK); Proyecto Aldea Global (Honduras); Merciphil Development Foundation (Philippines)	Offices in Azerbaijan, Bosnia-Herzegovina, Croatia, Guatemala, Kazakhstan, Kosovo, Macedonia, Kyrgyzstan, Lebanon, Nicaragua, Pakistan, Tajikistan, Uzbekistan; coverage in many other countries	

Name	Mission	Structure	Geographic Coverage	Revenue and Government Support
Oxfam International	Emergency and development projects; "work, when possible, with local organizations to rehabilitate communities . . . development projects are often small."[1]	International Secretariat (UK), Oxfam America, Oxfam-in-Belgium, Oxfam Canada, Community Aid Abroad (Australia), Oxfam GB, Oxfam Hong Kong, Intermon (Spain), Oxfam Ireland, Oxfam New Zealand, Novib (Netherlands), Oxfam Quebec		FY 1997: subscriptions from affiliates: $514,344; donations received: $462,052; other income: $26,074
Save the Children	To create safe places, enable family reunification and resettlement, provide land mine education, food security, and social and psychological assistance to children affected by war; Children in Crisis program; to provide basic education, economic assistance, and health care (Children in the developing world programs).]	Save the Children Fund Australia; Rettet das Kind, Austria; Save the Children Canada; Red Barnet, Denmark; Fundacion para el Desarrollo Comunitario, Dominican Republic, Egyptian Save the Children; Barnabati, Faroe Islands; Pelastakaa Lapset RY, Finland; Enfants et Developpement, France; Save the Children Greece; Alianza para el Desarrollo Juvenil Comuntario, Guatemala; Barnaheill, Iceland; Save the Children Japan; Jordanian Save the Children; Save the Children Korea; Save the Children of Macedonia; Save the Children Mauritius; Fundacion Mexicana de Apoyo Infantil AC, Mexico; Save the Children the	Worldwide	

Name	Mission	Structure	Geographic Coverage	Revenue and Government Support
Save the Children		Netherlands; Save the Children Fund New Zealand; Redd Barna, Norway; Salvati Copii, Romania; Rädda Barnen, Sweden; Save the Children Fund, UK; Save the Children Federation, USA; European Union Liaison office, Brussels	Worldwide	

Name	Function
International Council of Voluntary Agencies (ICVA)[k]	Provides a "permanent international liaison structure for voluntary agency consultation and cooperation . . . does not implement relief or development projects, but provides services and support to its member agencies to enable them to cooperate and perform more effectively."
	"Substantive work is done in Commissions that are established to respond to members' concerns in the areas of humanitarian assistance and development cooperation."
	Helps other NGOs establish networks.

[a] Action Against Hunger information pamphlet, n.d.

[b] ADRA International homepage: http://www.adra.org

[c] CARE USA homepage: http://www.care.org

[d] Caritas Internationalis homepage: http://www.caritasint.org

[e] Concern homepage: http://www.concern.ie/about_concern

[f] Doctors Without Borders USA homepage: http://www.dwb.org; international homepage: http://www.msf.org

[g] ICRS homepage: http://www.icrc.org

[h] Mercy Corps International homepage: http://www.mercycorps.org

[i] Oxfam International homepage: http://www.oxfaminternational.org

[j] U.S. Save the Children Federation homepage: http://www.savethechildren.org

[k] Umbrella organization.

FRENCH EXPERIENCE AND PERSPECTIVES

Although France has been especially active in West and Central Africa, its engagement is global. In terms of humanitarian airlift alone, France has engaged in some 70 relief operations in the period 1968–1998—roughly half in Africa, and the rest in Europe, Asia, and Latin America. French humanitarian airlift activity in Africa is first in rank among European allies; it is supported by permanent bases at Djibouti, N'Djamena, Libreville, Dakar, and Abidjan. In addition, Paris has defense arrangements with 23 African states.

HUMANITARIAN INTERVENTION AS A CORE MISSION

There are multiple reasons for French activism in relief operations beyond post-colonial links. Paris views humanitarian involvement in Africa as part of a larger vision of French leadership in francophone Africa. Humanitarian action is seen as part of the French foreign policy tradition, and expeditionary operations are very much part of French military tradition. Recent French governments, especially in the Mitterrand years, have also made an effort to transform the traditional pattern of French relations in Africa into a more multilateral strategy (e.g., working with the Organization of African Unity (OAU) and others to develop local capabilities for crisis management) emphasizing Third World-oriented development issues.[1] France has been among the strongest advocates of a *droit d'ingerence*—the "right to interfere" in humanitarian crises, and has combined vigorous humanitarian diplomacy in the UN and elsewhere with an as-

---

[1]Tiersky (1995), p. 51.

sertive approach to peacekeeping and peace support operations worldwide.[2] On a regional basis (e.g., in Africa), and in terms of its willingness to commit forces to complex, expeditionary operations, it is a "peer plus."[3]

French security strategy emphasizes humanitarian missions, alongside requirements for territorial defense, managing regional conflicts, maintaining defense agreements, and addressing a major threat in Europe. Humanitarian intervention is thus a core mission for French planners.[4] Moreover, French policymakers have generally been more willing than their American counterparts in recent years to use force in a limited, expeditionary manner and for crisis management—even where it has been difficult to articulate precise objectives.[5] The risk of "mission creep" and the necessity for "exit strategies" have been less evident in French debates on humanitarian intervention. French strategic culture imposes fewer constraints on the use of military forces for humanitarian purposes and places fewer obstacles to the withdrawal of these forces in circumstances short of "success" (often an intangible definition in relief operations). In short, the French political and operational calculus is more tolerant of murkiness in such operations.

French bases in and around Africa can greatly facilitate relief operations where the French military is involved (these facilities are not ordinarily available for French or other NGOs). The French airlift command (Commandement de la Force Aerienne de Projection—FAP) maintains permanent, prepositioned forces at four bases in the western hemisphere—three in Africa (Dakar, Fazsoi, and Djibouti) and at New Caledonia in the Pacific. In addition, there are now forces temporarily deployed in Chad, Abidjan, Gabon, N'Djamena, and, of course, the Balkans. The number and geographical distribution of French humanitarian airlift missions over the last 20 years is impressive.

---

[2]Guillot (1994), pp. 30–43.

[3]In recent years, French forces have engaged in "complex" interventions in the Central African Republic, Chad, Congo-Brazzaville, Rwanda, Zaire, and Somalia, along with many lesser deployments elsewhere.

[4]Lanxade (1994).

[5]For a good survey of changes in French defense strategy through the mid-1990s, see Laird (1995).

The mission emphases and operational challenges facing French airlift forces have evolved considerably since 1945, when the focus was on the evacuation of wounded personnel and the repatriation of prisoners and refugees. The environment at that time was unthreatening and did not call for specialized aircraft. In the 1960s, the focus shifted to disaster relief and food delivery, at longer range and sometimes to austere facilities, but again, with little threat. In the 1980s to the present, humanitarian airlift became engaged in a number of new and more complex missions—in insecure conditions.

NEW SECURITY RISKS

French planners emphasize that relief operations, especially in Africa (the same can be said of the Balkans) have become increasingly dangerous, to the extent that few missions are simply "humanitarian" in the strict sense. As a result, there are now greater demands on intelligence and self-protection ("force protection"). Airlift operations now often require the presence of armored vehicles at local airfields. With the growth of more serious SAM and air-to-air threats in relief operations, France has equipped its tactical airlifters with radar warning equipment and flares. Force protection requirements often dictate rapid loading and unloading of humanitarian cargo, placing a premium on short-takeoff, tactical aircraft (e.g., Transall). Tactical lift and short loiter times are also useful for non-combatant evacuations—an important mission for French airlifters. Heavy lift, largely the province of commercial vendors for Europeans, is judged to be less useful for relief operations of the sort France has been engaged in over the past decade. Its utility, in the French view, is largely confined to transit between main operating bases in Europe and Africa.

NGO-MILITARY RELATIONS

French NGOs and the military have a wary relationship, as is generally the case elsewhere. French embassies in host countries are the normal clearinghouse for liaison between NGOs and the French military. French NGOs are jealous of their independence, not only vis-à-vis the military but also in their relations with the French government. Ironically, this relationship was not improved by the elevation of humanitarian issues in French policy in recent years. Jealous of their autonomy and independence from the government, NGOs were particularly suspicious of repeated attempts to establish a Ministry of Humanitarian Affairs. The ministry had four separate incarnations and at one point was led by a founder of Médecins Sans Frontières.

Nonetheless, French NGOs have considerable respect for the professionalism and capabilities of the French military. NGOs and the military worked together in Ethiopia in 1985 and in Somalia, making good use of the French base in Djibouti. During Operation Turquoise in Rwanda in 1994, French NGOs worked alongside the French military for six months. Despite some suspicion, there was also a realization of considerable complementarity, especially in the medical area. This experience was repeated in French relief efforts in Honduras and Guatemala in the wake of Hurricane Mitch. While acknowledging the skills accumulated by the French in tropical medicine—another legacy of the colonial experience—the generation of military doctors with this experience is almost past. It is worth noting that the French military's assessment of the working relationship with NGOs is, in general, more negative than the assessment given by NGO officials.[6] The Programme Alimentaire Mondial (PAM) is cited by FAP as a positive exception in terms of capability and coordination with the military. Beyond the need for better coordination with NGOs, French airlifters note the more fundamental need for improved coordination *between* NGOs, and among NGOs, the Foreign Ministry, and the Ministry of Defense.

COOPERATION WITH OTHER EUROPEAN ALLIES

Germany, the UK, and, to a lesser extent, Belgium are the key European partners for France in humanitarian airlift. These working relationships are viewed as very good. Germany and France worked closely in Ethiopia in 1986 and have joint discussions (although no exercises) on their operations in Sudan. In some cases, operations have been multinational in a broader sense, as with Operation Pelican in Congo, where of the 6000 people evacuated by French aircraft only 2000 were French. There is an expectation that the interoperability with European allies will grow in light of such operations.

---

[6]An American observer with experience in Africa noted that the French military appeared "much tougher" in dealing with NGOs—including their own.

Interoperability will need to exist at the juridical level as well. As an example, there have been significant liability issues associated with Medivac cooperation with European partners. In coalition relief operations, the French military prefers to see a lead nation (e.g., France in Macedonia) responsible for the coordination of civil and military airlift. Traffic management is cited as having been a particular problem in Rwanda, where chartered aircraft often arrived without notice.

PERCEPTIONS OF THE U.S. ROLE IN COMPLEX RELIEF OPERATIONS

French military officials and French NGOs offer similar observations about U.S. military participation in humanitarian relief. The United States does things professionally, on a large scale, but is seen as "somewhat remote and disengaged." By contrast, French and other European militaries see themselves as knowing the countries and the cultures well, but necessarily doing things on a smaller scale.[7] The USAF is viewed as particularly effective at main operating bases and as having superb command, control, and communications (C3). French and American approaches to humanitarian intelligence also differ. The French and other Europeans tend to concentrate on local conditions—the situation on the ground. The United States in contrast, is in a better position to offer a regional overview (a well-known human intelligence versus technical means argument applied to the humanitarian environment). These differences offer the potential for considerable synergy; the French acknowledge that despite their considerable knowledge of the countries, military planners faced difficult intelligence problems in the Great Lakes crisis and had divergent indicators at critical points.

The French-U.S. working relationship at the operational level is described as very good. The Balkans offer a much more substantial test in this regard; in some previous cases, notably Somalia, airlifters from the two countries were engaged but operated from separate bases. The 1998 RECAMP exercise in Senegal brought together French, British, and U.S. personnel in a multilateral training activity

---

[7]For example, many of the Europeans involved in Somalia relief operations had already spent two years in Africa.

with African militaries. The emphasis was on peacekeeping, but the exercise was also relevant to humanitarian contingencies. France furnished aircraft and logistic support, while the U.S. contingent operated offshore.[8]

STANDARDIZATION IN HUMANITARIAN LIFT AND AIRDROP

The French airlift command sees considerable potential for improved standardization in airlift practices for humanitarian contingencies. The need for standardization is especially evident in airdrop, where allies have different practices. At the most basic level, pallets for airdrop may differ. Procedures for low-level airdrop, in particular, can also differ substantially. There is a need for uniform standards and, above all, exercises among allies to develop unified concepts for humanitarian airdrop (France has participated in some recent exercises of this sort with German and Polish forces). High-altitude airdrop (5000–15,000 feet), as practiced by U.S., French, and German forces in high-threat environments such as Bosnia, similarly requires special equipment and techniques. The FAP is developing new systems for this kind of high-altitude operation.

The range of techniques to suit varying humanitarian needs, threat environments, and even political objectives on the ground can be wide. In Sudan, French and German aircraft have been making low-altitude drops without pallets. During famine relief operations in the Sahel in the 1970s, France made airdrops at 60 feet without pallets— perhaps 50 percent of supplies dropped in this manner arrived intact.[9] The implications of these approaches for local politics and security conditions can be significant. For example, palletized drops tend to reinforce the local control of regimes, warlords, and armed factions, especially in poor weather or rugged terrain. Smaller, unpalletized packages can be more easily recovered by individuals.

---

[8]The British used their base at Ascension Island to support this exercise.

[9]German airlifters have made a specialty of dropping wooden pallets at low altitude without parachutes; French airdrops of this sort are almost always made with parachutes.

LESSONS FROM OPERATION PELICAN

The French evacuation and relief mission to Zaire and the Congo was conducted in three phases (Operations Pelican 1–3) between March and June 1997 and involved the movement of roughly 6000 people. Some 1500 were evacuated by civilian aircraft, and the remainder by military airlift. French planners have distilled several lessons from this experience. First, the operation confirmed the importance of tactical lift for the kinds of missions France is likely to face in the future. Second, the operation reinforced the perception that such missions are increasingly "complex," with political and security implications that transcend the traditional definition of humanitarian intervention. This complexity, including force protection problems, also made cooperation and coordination between air and ground elements critical. Third, the Pelican experience underscored the value of prepositioning aircraft, personnel, and equipment in areas of likely demand, and thus the value of France's network of African bases.

FRANCE AS A KEY PARTNER

The high level of French civilian and military engagement in the worldwide humanitarian sphere, together with a relevant overseas basing structure and significant tactical lift assets, some forward deployed, make France a key partner for coalition operations. In many cases, France will be a lead state for European or transatlantic efforts. Moreover, like the United States, what France does with regional militaries and other allies will have a strong influence on the capacity for local response in complex relief operations, especially in Africa. Finally, there is strong professional interest among French planners and operators in developing closer military-to-military cooperation with the United States on support of relief operations. Because France is likely to be a strong force behind future EU defense efforts, including expeditionary capabilities, more effective cooperation with France at the political and operational levels can translate into a more effective overall partnership with European allies in managing humanitarian crises.

BIBLIOGRAPHY

Action Against Hunger, *Geopolitics of Hunger 1998–1999, Using Hunger as a Weapon*, London, 1999 (originally published as *Géopolitique de la faim*, Presses Universitaires de France, Paris, 1998).

Allard, Kenneth, *Somalia Operations: Lessons Learned*, National Defense University, Washington, DC, 1995.

Associated Press, "Nations Pledging to Join the Force," *New York Times*, September 16, 1999.

Balanzino, Sergio, "NATO's Humanitarian Support to the Victims of the Kosovo Crisis," *NATO Review*, Summer 1999.

"The Balkan Refugee Crisis: Regional and Long-term Perspectives," downloaded from http://wwwnotes.reliefweb.int/files/rwdo, released June 2, 1999, downloaded June 16, 1999.

Barber, Ben, "Feeding Refugees, or War?" *Foreign Affairs*, Vol. 76, No. 4 July/August 1997, pp. 8–14.

Barnes, Rudolph C., Jr., "Civic Action, Humanitarian and Civic Assistance, and Disaster Relief: Military Priorities in Low-intensity Conflict," *Special Warfare*, Vol. 2, No. 4, Fall 1989, pp. 34–41.

Baumann, Robert F., "Operation Uphold Democracy: Power Under Control." *Military Review*, July–August 1997, downloaded from http://www-cgsc.army.mil/milreview/english/julaug97/baumann.htm, June 1, 1999.

Becker, Elizabeth, "With NATO in Charge, Relief Looks Less Neutral," *New York Times*, April 10, 1999, p. A8.

Bernstein, Joanne, "Military Assistance in Sub-Saharan Africa," *The DISAM Journal*, Fall 1994, pp. 90–103.

Betts, Richard K, "The Delusion of Impartial Intervention," *Foreign Affairs*, Vol. 73, No. 6, November/December 1994, pp. 20–33.

Boutroue, Joel, "Missed Opportunities: The Role of the International Community in the Return of the Rwandan Refugees from Eastern Zaire," *The Rosemary Rogers Working Paper Series*, Working Paper #1, June 1998.

Bowden, Mark, *Black Hawk Down: A Story of Modern War*, Atlantic Monthly Press, Boston, MA, 1999.

Brauman, Rony, in Kevin Cahill (ed.), *A Framework for Survival, Health, Human Rights and Humanitarian Assistance in Conflicts and Disasters*, Basic Books, New York, 1993.

Brauman, Rony, and Joelle Tanguy, *The Médecins Sans Frontières Experience*, Doctors Without Borders, New York, 1998.

Byman, Daniel, and Stephen Van Evera, "Why They Fight: Hypotheses on the Causes of Contemporary Deadly Conflict," *Security Studies*, Spring 1998, pp. 1 50.

Carnegie Commission on Preventing Deadly Conflict, *Preventing Deadly Conflict: Final Report*, Carnegie Foundation, Washington, DC, 1997.

Catholic Medical Mission Board, *Medical Mission News*, Vol. 68, No. 4, New York, Winter 1998.

Center of Excellence in Disaster Management & Humanitarian Assistance, *Annual Report 1998*, Tripler Army Medical Center, Hawaii, 1998.

Center of Excellence in Disaster Management & Humanitarian Assistance, *Annual Report 1999*, Tripler Army Medical Center, Hawaii, 1999.

Center of Excellence in Disaster Management & Humanitarian Assistance, *Brave Knight 1998, 26 February–3 March 1998 After Action Report*, Tripler Army Medical Center, Hawaii, 1999.

Center of Excellence in Disaster Management & Humanitarian Assistance, *Combined Humanitarian Assistance Response Training*, Tripler Army Medical Center, Hawaii, 1999.

Center of Excellence in Disaster Management & Humanitarian Assistance, *Health Emergencies in Large Populations Course, Major Participants in Humanitarian Assistance*, Tripler Army Medical Center, Hawaii, 1999.

Clark, Ann Marie, Elisabeth Friedman, and Kathryn Hochstetler, "The Sovereign Limits of World Civil Society," *World Politics*, Vol. 51, October 1998.

Cohen, Roberta, and Francis M. Deng, *Masses in Flight, The Global Crisis of Internal Displacement*, Brookings Institution Press, Washington, DC, 1998.

Cooperation in Peacekeeping: Workshop on Humanitarian Aspects of Peacekeeping, NATO (99)16, Geneva, February 10–12, 1999.

Department of Defense Dictionary of Military and Associated Terms, Diane Publishing, Upland, PA, 1997.

Duffield, Mark, and John Prendergast, *Without Troops and Tanks: Humanitarian Intervention in Ethiopia and Eritrea*, Red Sea Press, Lawrenceville, NJ, 1994.

Dworken, Jonathan, "Coordinating Relief Operations," *Joint Forces Quarterly*, Summer 1995.

Dworken, Jonathan T., *Improving Marine Coordination with Relief Organizations in Humanitarian Assistance Operations*, Center for Naval Analysis, Alexandria, VA, 1996.

Forman, Shepard, and Rita Parhad, "Paying for Essentials: Resources for Humanitarian Assistance," *Journal of Humanitarian Assistance*, http://www-jha.sps.cam.ac.uk/a/a404htm, posted December 14, 1997.

Geoghegan, Tracy, and Kristen Allen (eds.), *InterAction Member Profiles, 1997–1998*, American Council for Voluntary International Action, Washington, DC, 1997.

Gourevitch, Philip, *We Wish to Inform You That Tomorrow We Will Be Killed with Our Families: Stories from Rwanda*, Farrar Straus and Giroux, New York, 1998.

Gow, James, *Triumph of the Lack of Will, International Diplomacy and the Yugoslav War*, Columbia University Press, New York, 1997.

Guillot, Philippe, "France, Peacekeeping and Humanitarian Intervention," *International Peacekeeping*, Vol. 1, No. 1, Spring 1994.

"Guns or Refugees—an Unequal Alliance?" *The Economist*, May 8, 1999, p. 50.

Hirsch, John L., and Robert B. Oakley, *Somalia and Operation Restore Hope, Reflections of Peacemaking and Peacekeeping*, United States Institute of Peace Press, Washington, DC, 1995.

Houston, A. M., "Comments on AMC Operations During Support Hope," Memorandum, United Nations High Commissioner for Refugees, Geneva, Switzerland, 1994.

International Fund for Agricultural Development (IFAD), *The State of Rural Poverty: A Profile of Africa*, IFAD, Rome, 1993.

Joint Chiefs of Staff, *Joint Military Doctrine for Military Operations Other Than War*, Joint Publication 3-07, Washington, DC, June 16, 1996.

Joint Chiefs of Staff, *Interagency Coordination During Joint Operations*, Joint Publication 3-08, Vol. II, Washington, DC, October 9, 1996.

Joint Chiefs of Staff, *Joint Tactics, Techniques, and Procedures for Peace Operations*, Joint Publication 3-07.3, Washington, DC, February 12, 1999.

Joulwan, George A., and Christopher C. Shoemaker, *Civilian-Military Cooperation in the Prevention of Deadly Conflict*, Carnegie Corporation, New York, 1999.

Kaminski, Matthew, and A. Craig Copetas, "Nations Are Divided on Method of Handling Kosovo Refugees," *Wall Street Journal*, April 7, 1999.

Keen, D., and K. Wilson, "Engaging with Violence: Relief in Wartime," *War and Hunger: Rethinking International Responses to Complex Emergencies*, Zed, London, 1994.

Kennedy, Kevin M., "The Relationship Between the Military and Humanitarian Organizations in Operation Restore Hope," in Walter Clarke and Jeffrey Herbst (eds.), *Learning from Somalia: The Lessons of Armed Humanitarian Intervention*, Westview Press, Boulder, CO, 1997, pp. 99–117.

Khalilzad, Zalmay, and Ian O. Lesser (eds.), *Sources of Conflict in the 21st Century, Regional Futures and U.S. Strategy*, RAND, MR-897-AF, 1998.

Klaiber, Klaus-Peter, *Chairman's Summary: Workshop on Humanitarian Aspects of Peacekeeping*, EAPC (PMSC-AHG)N(99)11, Geneva, February 10–12, 1999.

Knight, Tim, *Bosnia-Herzegovina: Basic Themes from a Complex Emergency*, Feinstein International Famine Center, Tufts University, Medford, MA, June 1998.

Kozaryn, Linda D., "NATO Approves Kosovo Plans," *Armed Forces Press Service*, August 14, 1998; downloaded from http://www.defenselink.mil/news/Aug1998, June 24, 1999.

Kuhne, Winrich, Guido Lenzi, and Alvaro Vasconcelos, "WEU's Role in Crisis Management and Conflict Resolution in Sub-Saharan Africa," *Chaillot Papers*, No. 22, WEU Institute for Security Studies, Paris, December 1995.

Laird, Robbin, "French Security Policy in Transition: Dynamics of Continuity and Change," *McNair Papers*, No. 38, Institute for National Strategic Studies/National Defense University (INSS/NDU), March 1995.

Landay, Jonathan S., and Peter Ford, "Why Aid Workers Call Kosovo the Toughest Case," *Christian Science Monitor*, May 12, 1999;

downloaded from http://www.csmonitor.com/durable/1999/05/12/p9s1.htm, June 6, 1999.

Lanxade, Admiral Jacques, "French Defense Policy After the White Paper," *RUSI Journal*, April 1994.

Laurence, Edward J., "Light Weapons and Intrastate Conflict: Early Warning Factors and Preventative Action," Carnegie Commission on Preventing Deadly Conflict, July 1998; downloaded from http://www.ccpdc.org/pubs/weapons.htm, April 20, 1999.

Lesser, Ian O., Jerrold Green, F. Stephen Larrabee, and Michele Zanini, *The Future of NATO's Mediterranean Initiative: Evolution and Next Steps*, RAND, MR-1164-SMD, 1999.

Lightburn, David, "NATO and the Challenge of Multi-Functional Peacekeeping," *NATO Review*, March 1996.

Medical Mission News, Winter 1998.

Meier, Barry, "Supplies-Side Economics: When Disaster Strikes, Someone Must Provide the Tents," *New York Times*, May 5, 1999, p. C1.

Mendiluce, Jose Maria, "Meeting the Challenge of Refugees: Growing Cooperation Between UNHCR and NATO," *NATO Review*, April 1994.

Menkhaus, Ken, "Complex Emergencies, Humanitarianism, and National Security." *National Security Studies Quarterly*, Vol. IV, Issue 4, Autumn 1998, pp. 53–61.

Mets, David R., *Land-Based Air Power in Third World Crises*, Air University Press, Maxwell Air Force Base, AL, 1986.

Miller, Judith, "UN's Workers Become Targets in Angry Lands," *New York Times*, September 19, 1999.

Natsios, Andrew S., "The International Humanitarian Response System," *Parameters*, Spring 1995.

Natsios, Andrew S., "Humanitarian Relief Intervention in Somalia: The Economics of Chaos," in Walter Clarke and Jeffrey Herbst (eds.), *Learning from Somalia: The Lesson of Armed Humanitarian Intervention*, Westview, Boulder, CO, 1997.

Natsios, Andrew S., *U.S. Foreign Policy and the Four Horsemen of the Apocalypse*, Praeger, Westport, CT, 1997.

Newett, Sandra L., *Planning for Humanitarian Assistance Operations*, Center for Naval Analysis, Alexandria, VA, April 1996.

"The Non-Governmental Order," *The Economist*, December 11, 1999, pp. 20–21.

North Atlantic Council, The Washington Declaration signed by Heads of State participating in meeting of the North Atlantic Council, NAC-S(99)63, Washington, DC, April 23, 1999.

North Atlantic Council, The Alliance's Strategic Concept approved by Heads of State participating in meeting of the North Atlantic Council, NAC-S(99)65, Washington, DC, April 24, 1999.

Oakley, Robert B., Michael J. Dziedzic, and Eliot M. Goldberg, *Policing the New World Disorder: Peace Operations and Public Security*, National Defense University Press, Washington, DC, 1998.

Palmeri, Francesco, "Civil Emergency Planning: A Valuable Form of Cooperation Emerges from the Shadows," *NATO Review*, No. 2, 1996.

Palmeri, Francesco, "A Euro-Atlantic Disaster Response Capability," *NATO Review*, Autumn 1998.

Pirnie, Bruce R., *Civilians and Soldiers, Achieving Better Co-ordination*, RAND, MR-1026-SRF, 1998.

Pirnie, Bruce, and Corazon M. Francisco, *Assessing Requirements for Peacekeeping, Humanitarian Assistance, and Disaster Relief*, RAND, MR-951-OSD, 1998.

Pope, Hugh, "Aid Agencies Adopt Ways of Business as Competition Grows to Help Kosovo," *Wall Street Journal*, June 18, 1999 (electronic version).

Rowland, Wade, *The Plot to Save the World*, Clarke, Irwin & Co., Toronto, 1973.

Rupiah, Lt. Colonel Martin (ret.), "Peacekeeping Operations: The Zimbabwean Experience," *Journal of Humanitarian Assistance*; from http://www-jha.sps.cam.ac.uk/a/a061.htm, posted 25 January 1996.

Scarborough, Rowan, "Study Hits White House on Peacekeeping Missions," *Washington Times*, December 6, 1999, p. 1.

Schoettle, Enid, "Information Sharing in Humanitarian Emergencies: Progress and Problems," talk given at Meridian House workshop sponsored by the Defense Intelligence Agency, the National Defense University, the Office of the Secretary of Defense, and InterAction, June 29, 1998.

Seiple, Chris, *The U.S. Military/NGO Relationship in Humanitarian Interventions*, U.S. Army War College, Peacekeeping Institute, Carlisle, PA, 1996.

Senior NATO Logisticians' Conference Secretariat, *NATO Logistics Handbook*, NATO, Brussels, 1997.

Seybolt, Taylor B., "The Myth of Neutrality," *Peace Review*, Vol. 8, No. 4, 1996, pp. 521–527.

Siegel, Adam B., *Requirements for Humanitarian Assistance and Peace Operations: Insights for Seven Case Studies*, Center for Naval Analysis, Alexandria, VA, March 1995.

"Sins of the Secular Missionaries," *The Economist*, January 29, 2000, pp. 25–27.

Slim, Hugo, "Planning Between Danger and Opportunity: NGO Situation Analysis in Conflict Related Emergencies," *Journal of Humanitarian Assistance*, January 16, 1996.

Stewart, Robert A., *Broken Lives, A Personal View of the Bosnian Conflict*, HarperCollins, New York, 1993.

Taft, Julia, and James Pardew, "Transcript on Aid to Kosovo Refugees." November 13, 1998; downloaded from http://www.

usia.gov/regional/eur/balkans/kosovo/texts/1113taft.htm, June 6, 1999.

Taw, Jennifer M., *Operation Just Cause: Lessons for Operations Other Than War*, RAND, MR-569-A, 1996.

Terrel, David L., *Airlift Operations Study, United Nations High Commissioner for Refugees (UNHCR)*, United States European Command, Stuutgart, Germany, 1994.

Thompson, Keith, "Relief Effort Transforms Sleepy Terminal to Humanitarian Hub," http://www.af.mil, released 4 May, 1999, downloaded June 9, 1999.

Tiersky, Ronald, "The Mitterrand Legacy and the Future of French Security Policy," *McNair Papers*, No. 43, Institute for National Strategic Studies/National Defense University (INSS/NDU), Washington, DC, August 1995.

United Nations, *The Blue Helmets, A Review of United Nations Peace-Keeping*, Third Edition, United Nations Department of Public Information, New York, 1996.

United Nations, *Basic Facts About the United Nations*, New York, 1998.

United Nations Department of Humanitarian Affairs, *The Use of Military and Civil Defence Assets in Disaster Relief Operations, MCDA Field Manual*, United Nations, New York, 1996.

United Nations High Commissioner for Refugees, *Review of UNHCR Logistics Policies and Practices*, Central Evaluation Section, EVAL/LOG/12, Geneva, 1992.

United Nations High Commissioner for Refugees, "Emergency," *Refugees*, No. 103, Geneva, February 1993.

United Nations High Commissioner for Refugees, *Human Rights*, No. 92, Geneva, April 1993.

United Nations High Commissioner for Refugees, *The High Cost of Caring*, No. 102, Geneva, 1995.

United Nations High Commissioner for Refugees, Survey, *Working with the Military*, UNHCR, Geneva, 1995.

United Nations High Commissioner for Refugees, *The State of the World's Refugees*, Oxford University Press, New York, 1997.

United Nations High Commission for Human Rights, "Report by the High Commission for Human Rights on the Situation of Human Rights in Kosovo," May 31, 1999; downloaded from http://www.reliefweb.int, June 17, 1999.

United Nations High Commissioner for Refugees and Office of the Coordination of Humanitarian Assistance, "UN Seeks US $54.3 million for Kosovo," September 8, 1998, downloaded from http://www.unhcr.cn/news/pr/pr980909.htm, June 25, 1999.

United Nations High Commissioner for Refugees, "1999 Global Appeal," downloaded from http://www.unhcr.ch.fdrs/ga99/mkd.htm, June 21, 1999.

United Nations High Commissioner for Refugees, "UNHCR by Numbers," http://www.unhcr.cn/un&ref/numbers.

United States European Command, *Operation Support Hope: After Action Review* (no date).

U.S. Government, *The Clinton Administration's Policy on Managing Complex Contingency Operations*, Presidential Decision Directive 56, White Paper, Washington DC, May 1997.

Vick, Alan, David T. Orletsky, Abram N. Shulsky, and John Stillion, *Preparing the U.S. Air Force for Operations Other Than War*, RAND, MR-842-AF, 1997.

de Vries, Fokko, "A Doctor's Diary," http://www.msf.org/project/yugoslavia/kosovo/testimony/1999/04/diary, posted on April 20, 1999, accessed June 15, 1999.

Wallensteen, Peter, "Armed Conflict and Regional Conflict Complexes, 1989–97," *Journal of Peace Research*, Vol. 35, No. 5, 1998, pp. 621–634.

Wentz, Larry K. (ed.), *Lessons Learned from Bosnia: The IFOR Experience*, C4ISR Cooperative Research Program, Washington, DC, 1998.

Wiener, Michael, with Nitza Machmias, "To Save One Life Is to Save the World: The Israeli Operation in Rwanda," in Eric A. Belgrad and Nitza Machmias (eds.), *The Politics of International Humanitarian Air Operations*, Praeger, Westport, CT, 1997.

Woodward, Susan L., *Balkan Tragedy: Chaos and Dissolution After the Cold War*, Brookings Institution, Washington, DC, 1995.

Zandee, Dick, "Civil-Military Interaction in Peace Operations," *NATO Review*, Spring 1999.